TRANSLATIONAL SYSTEMS BIOLOGY

TRANSLATIONAL SYSTEMS BIOLOGY

CONCEPTS AND PRACTICE FOR THE FUTURE OF BIOMEDICAL RESEARCH

YORAM VODOVOTZ

*Department of Surgery and McGowan Institute
for Regenerative Medicine, University of Pittsburgh, Pittsburgh, PA, USA*

GARY AN

Department of Surgery, University of Chicago, Chicago, IL, USA

AMSTERDAM • BOSTON • HEIDELBERG • LONDON • NEW YORK • OXFORD
PARIS • SAN DIEGO • SAN FRANCISCO • SINGAPORE • SYDNEY • TOKYO
Academic Press is an imprint of Elsevier

Academic Press is an imprint of Elsevier
32 Jamestown Road, London NW1 7BY, UK
525 B Street, Suite 1800, San Diego, CA 92101-4495, USA
225 Wyman Street, Waltham, MA 02451, USA
The Boulevard, Langford Lane, Kidlington, Oxford OX5 1GB, UK

ISBN: 978-0-12-397884-4

British Library Cataloguing-in-Publication Data
A catalogue record for this book is available from the British Library

Library of Congress Cataloging-in-Publication Data
A catalog record for this book is available from the Library of Congress

For Information on all Academic Press publications
visit our website at http://store.elsevier.com/

Typeset by MPS Limited, Chennai, India
www.adi-mps.com

Working together
to grow libraries in
developing countries

www.elsevier.com • www.bookaid.org

Dedication

This book is dedicated to our families—Melanie, Madeline, Xing,
Lena, and Ethan—who have stood by us, tolerated our rants, acted as our sounding boards,
and provided relief and grounding.

Contents

III TRANSLATIONAL SYSTEMS BIOLOGY: HOW WE PROPOSE TO FIX THE PROBLEMS OF THE CURRENT BIOMEDICAL RESEARCH LANDSCAPE

IV TOOLS AND IMPLEMENTATION OF TRANSLATIONAL SYSTEMS BIOLOGY: THIS IS HOW WE DO IT

Preface

The more pity that fools may not speak wisely what wise men do foolishly. **Touchstone, As You Like It, Act 1, Scene 2, Line 85**

Truth's a dog must to kennel; he must be whipp'd out... **The Fool, King Lear, Act 1, Scene 4, Lines 109–110**

Foolery, sir, does walk about the orb, like the sun; it shines everywhere. **Feste, Twelfth Night, Act 3, Scene 1, Lines 37–38**

The fool doth think he is wise, but the wise man knows himself to be a fool. **Touchstone, As You Like It, Act 5, Scene 1, Lines 31–32**

Some of Shakespeare's most memorable characters are Fools. Whether in Comedy or Tragedy, the Fools in Shakespeare's plays serve an important role that mirrored their historical position. Whether by intellect, simplicity, or station, they embody the "other" in Shakespeare's worlds, standing outside the primary context of the play and thereby serving to provide an "honest" commentary on the proceedings, stripped of the social niceties and motivations that temper the comments of the other characters. Their position as Fools to a great degree protects them from the consequences of their statements and actions. In the historical (?) eras of Absolutism, Fools often provided the only means of vocalizing critiques of rulers and their societies; many scholars interpret their role as a vital social mechanism for funneling and focusing dissent in a nondisruptive and nonthreatening way. A Fool might make the comments regarding a Royal personage's peccadillos that are on the forefront of the court members' minds but about which are unable to note or speak of for fear of repercussion; however, spouting forth from the mouth of the Fool these exact same thoughts can safely be greeted with laughter (albeit a bit ostentatiously nervous and not too long or loud...).

Fools are intrinsically paradoxical. They are able to express the most cutting of critiques, but in so doing reinforce the stability of the societies they critique. They are outsiders, but often with the most intimate of relationships with their masters (i.e., Lear and his Fool, Viola and Feste, Rosalind and Touchstone). This paradoxical duality is even manifest in the most recognizable guise of the Fool: the harlequin costume of opposing and alternating color blocks. In Shakespeare's plays, his most notable Fools, such as Touchstone, Feste, and Lear's Fool, are often seen as mouthpieces for the plays' author. While this interpretation remains the subject of many an English Major's thesis and open to scholarly debate, there is something intrinsically consistent with the paradoxical nature of Fools in the belief that the author, who cannot be more intimately related to the piece he is writing, places his own truest words in the mouth of the plays' representative outsider. Regardless of whether they represent Shakespeare's opinions and beliefs, certainly these three Fools appear to possess a perspective more encompassing than the other characters in the plays they inhabit. Touchstone is the voice of logical reasoning and objective assessment among the refugees in the Forest of Arden; Lear's Fool provides the most comprehensive view and assessment of the madness and chaos following Lear's fateful decision; and Feste appears at sometimes an omniscient observer regarding the machinations taking place in the Duchy of Illyria.

Perhaps, the lesson to be taken from Shakespeare's Fools is that possession of the wider perspective, a grasping of the big picture, is what provides the most uncomfortable and disruptive conclusions, which, for the sake of the society, can only be spoken by Fools. It is truly ironic, then, that at the end of each of the plays noted above, there has been a dramatic disruption and reshuffling of the conditions at their respective beginnings, those same conditions that required the placement of disruptive comments in a Fool's mouth in an attempt to forestall the paradigmatic shifts that eventually would come to pass. Perhaps, then, the Fool is not such a fool after all.

We contend that the biomedical research community could potentially benefit from listening to a pair of Fools. Today, in biomedical research, paradoxes abound. We know more about the generative processes and mechanisms of disease than at any other point in history. The rate of this data acquisition shows no signs of slowing down, yet the introduction of new and more effective therapeutics for the diseases that most trouble us has never been less efficient. Many of those diseases that most trouble us—cancer, sepsis, obesity, autoimmune disorders—represent the hijacking of otherwise beneficial biological processes: cancer of growth and healing, sepsis of inflammation, obesity of metabolism, and autoimmune disorders of immune protection. Even those therapeutics that have been life saving in the past (and still are in the present) have been revealed as producing a whole new set of problems that, in many ways, arise because of their initial success: antibiotics leading to increasingly lethal resistance, life support measures in the intensive care unit leading to the purgatory of chronic organ insufficiency and recurrent infections. Inside the biomedical community (for legitimate reasons we will discuss in this book), it is too easy to look only toward the patch of blue sky representing future technological and intellectual achievement. Yet, if one looks more widely (and not even much more widely, as the examples above can attest to) there are signs of the storm clouds building. This book is intended to help us turn around and look.

Contrary to genius, which is born, Fools might be made, through the correct and fortuitous mix of temperament, training, timing, and terrain. Two decades ago, the intellectual terrain of critical illness was ripe for disruption: clinical trials for the treatment of sepsis that, by all ostensible criteria for success, should have worked, did not. There was the beginning of an existential crisis in the biomedical research community that persists to this day in the Translational Dilemma. The traditional community would quickly retrench and retool, but for those with the right level of training (i.e., not too deeply embedded in the traditional academic paths) and the appropriate temperament (i.e., curious, contrarian, and stubborn about it), this was an opportune time to learn to become a Fool. About 15 years ago, working separately and initially without knowledge of each other's work, two very different people with very different backgrounds and for very different reasons began to question the state of biomedicine in the context of critical illness. Gary, a trauma surgeon and intensive care physician, saw a clinical dilemma, and being outside the academic research structure, was unburdened by knowledge and possessing of procedural naiveté. As such, he started on his own, on the fringe, with the naïve concept that if he, as someone without formal research training, could apply these new techniques to the challenges facing the translation of basic research knowledge, then surely just that demonstration would be enough to lead to adoption (after all, what could possibly go wrong!). Yoram, a researcher, a biochemist and immunologist, had experienced the academic biomedical research community at near its highest level, but with enough insight and perspective to recognize both its best and worst characteristics. He, too, pursued an unorthodox path, forming part of the leadership of an interdisciplinary team before that description had become *de riguer*, and enlisting the potential of industry by helping to start a biosimulation company. This odd couple had absolutely no reason to ever meet, to find common ground, to have compatible personalities, much less to collaborate. But, in 2002, they, as they say, "met cute" by total coincidence at lunch at the 25th Annual Meeting of the Shock Society. Conversation ensued, connections and common interests were identified, and a collaboration was born. Time and terrain contributed to foster the conditions that would enhance this collaboration, as at that 2002 meeting, Gary and Yoram were invited, along with their collaborators, to a workshop in Germany on complex systems approaches to critical illness. This led to the formation of the Society for Complex Acute Illness, which has and continues to serve as the focal point for the most dynamic discussions and developments in this still-growing field.

The simple fact that pursuing this line of investigation required the formation of a new scientific society points to the "outsider" nature of the endeavor. Yet while they remained outsiders, they kept hearing from many people that they were on the right track (perhaps the surreptitious support of the members of court that have an inkling that these Fools spoke some truth?). Self and system analysis ensued, and the germ of an idea grew out many late night discussions: what was it that they were doing that resonated with individual people but was distrusted, misunderstood, or ignored by so many others primarily at an organizational level? Were the problems, deficiencies, successes, and failures they observed specific to critical illness or a more general phenomenon?

Translational Systems Biology was the product of these discussions, disseminated since through multiple manuscripts on the topic, but facing a fair amount of resistance and indifference (after all, how dare the Fools try to be members of the court!). Hence the paradox of being the Fool, able to speak certain truths, but only given the freedom to do so because their voices serve to preserve the *status quo*. But, as seen in King Lear, Twelfth Night, and As You Like It, change will come. When the offer to write this book came along, we viewed this as perhaps a qualitative shift in how our message could be delivered, and hence, how it would be received. Few Fools are given this opportunity, to potentially have their words actually have real and tangible meaning. When exercising that opportunity, we realized that there needed to be a balance between tempering the traditional level of agitation that is the domain of Fools, but retaining the disruptive quality of the Foolery that first brought it attention. As such, readers should consider themselves warned that they may be rattled, that our chosen tone might bite, and that the images cast by the mirror provided might not be ones they would prefer to see. If, at some point, the reader might feel too accused, they should perhaps take at least some comfort in that, as all Fools are, the authors are part of the same system they are critiquing, and their reflections are present in that same mirror.

No one would subject themselves to that degree of uncomplimentary self-reflection unless it was for a hope that somehow doing so would make things better. It is all too easy to critique without providing an alternative solution.

We believe that we have avoided this trap: this is why we present Translational Systems Biology as a potential pathway to address what we interpret as the roadblocks and barriers to enhancing human health for the remainder of the twenty-first century and hopefully beyond. Consistent with our strategy of founding things on the fundamental principles of Science, we do not claim that this is a unique or even necessarily effective solution. Rather, we claim, that given a transparent analysis of the preponderance of the evidence, a problem with the structure and operations of the biomedical research community has been identified and we have proposed a *plausible* solution to that problem. If the result of this book is the establishment of new, competing views of how we as a research community can move forward, and these and our strategies are given the opportunity to be given the test of real-world implementation, then we can consider our endeavor a success. This, after all, is how Science progresses.

This book includes an *Acknowledgments* section, but we would like to take the last space of our *Preface* to give special thanks to those who most suffered the company of Fools, our families. Gary wishes to thank his lovely wife Melanie, who always wondered where this time consuming, semiprofessional hobby would lead, and his daughter Madeline, who may be cursed with too many similarities to her father; and Yoram wishes to thank his better half, Xing, who stood by his side and provided grounding, and his daughter Lena and his son Ethan, who grew up surrounded by the good and bad of Science.

Acknowledgments

The authors would like to acknowledge the contributions to this work of a large number of investigators, students, and postdoctoral fellows, with whom they have shared a career in science and without which this book could not have been written: Andrew Abboud, Khalid Almahmoud, John Alverdy, Derek Angus, Julia Arciero, Nabil Azhar, Steve Badylak, Derek Barclay, Arie Baratt, John Bartels, Binnie Betten, Bibiana Bielekova, Timothy R. Billiar, David Brienza, David Brown, Cliff Brubaker, Tim Buchman, Marius Buliga, Frederick D. Busche, David Carney, Steve Chang, Carson Chow, Scott Christley, Gilles Clermont, R. Chase Cockrell, Greg Constantine, Marie Csete, Judy Day, Edwin Dietch, Russell Delude, John Doyle, Joyeeta Dutta-Moscato, Bard Ermentrout, James Faeder, Jie Fan, Rena Feinman, Ira Fox, Ali Ghuma, Mitchell P. Fink, David Hackam, Pat Hebda, C. Anthony Hunt, Jelena Janjic, John Kellum, Moses Kim, Christine Kretz, Shilpa Krishnan, Swati Kulkarni, Rukmini Kumar, Claudio Lagoa, Ryan M. Levy, Nicole Li, Shirley Luckhart, Othman Malak, Qi Mi, Maxim Mikheev, John Murphy, Rajaie Namas, Rami Namas, Carl Nathan, Eddy Neugebauer, Gary Nieman, Juan Ochoa, David Okonkwo, Patricio Polanco, John Pollock, Ian Price, Jose M. Prince, Juan Carlos Puyana, Heinz Redl, Angela Reynolds, Beatrice Riviere, Glen Ropella, Matthew Rosengart, Jonathan Rubin, Alan Russell, David Sadowsky, Joydeep Sarkar, John Seal, Jason Sperry, Michael Sporn, Alexey Solovyev, Robert Squires, David L. Steed, Jordan Stern, Joshua Sullivan, David Swigon, Shlomo Ta'asan, Andres Torres, Jeffrey Upperman, Katherine Verdolini, Bill Wagner, Matt Wolf, Jinling Yin, Ivan Yotov, Akram Zaaqoq, Ruben Zamora, Cordelia Ziraldo, and Sven Zenker. The authors would also like to thank all of the students that participated over the years in the University of Pittsburgh's Systems Approach to Inflammation graduate course. Much of the inspiration we have received for our work over the years, and the inspiration and feedback we have gotten for the concepts embodied in this book, came from great friends who are not scientists: Michelangelo Celli, Renee Colbert, Ted Christie, Mike Farrell, Michael Greenberg, Ariel Kuperminc, Michael Martin, Meipo Fun-Martin, Scott Osterrieder, Barry Strubel, Clyde Takeguchi, and Nancy Wolper. The authors are especially grateful for the support for their work from the National Institutes of Health, the Department of Defense, the National Institute on Disability and Rehabilitation Research, the Commonwealth of Pennsylvania, the Pittsburgh Lifesciences Greenhouse, the Pittsburgh Tissue Engineering Initiative, and IBM, Inc.

INTRODUCTION AND OVERVIEW

1.1

Interesting Times: The Translational Dilemma and the Need for Translational Systems Biology of Inflammation

It was the best of times, it was the worst of times, it was the age of wisdom, it was the age of foolishness, it was the epoch of belief, it was the epoch of incredulity, it was the season of Light, it was the season of Darkness, it was the spring of hope, it was the winter of despair... **A Tale of Two Cities, Charles Dickens**

Consider the following scenarios:

You are driving home from a party and a drunk driver runs a red light, striking your car from the side, crushing the door and trapping you inside. The paramedics and firemen arrive quickly and cut you out, but you have lost a lot of blood and are in shock. They get you to the hospital, where you are found to have a broken leg, a broken pelvis, a collapsed and bruised lung, and are bleeding internally. You get a series of operations to stop the bleeding and fix the fractures *but end up in the Intensive Care Unit on a ventilator because your lungs were too badly damaged...*

You are recovering from a cycle of chemotherapy for your breast cancer, and your doctors are saying that you appear to be responding well, but a few days afterward you start having fevers and feeling very poorly. You call 911, and by the time the paramedics can get you to the hospital you have a very low blood pressure and are having trouble breathing. The Emergency Room doctors diagnose you with bacterial sepsis, because you are immunosuppressed from your chemotherapy. *They start intravenous fluids and antibiotics and transfer you to the Intensive Care Unit...*

It is flu season, and despite being careful you have come down with a bad cough. You stay home, drink fluids and have soup, but after about 4 days you start coughing up greenish-yellow phlegm and are sweating at night and have chills. Your family brings you to the hospital, where they say you have a pneumonia. You are admitted and placed on antibiotics, but over the next day your breathing becomes more difficult and your blood pressure starts to drop. *They transfer you to the Intensive Care Unit and tell you that then need to put you on a ventilator...*

You have been shoveling snow and have developed really bad chest pain. You call 911, and the paramedics take you to the emergency room where they diagnose you with a heart attack from occlusions in your heart's arteries. Based on where the blockage is, you need to have emergency heart bypass surgery. The operation goes fine, but afterward your kidneys no longer work so well and the wound on your leg where they took the vein graft is looking a bit red and maybe infected. After about 5 days you are having more trouble breathing and your *doctors say you need dialysis and transfer you back to the Intensive Care Unit...*

You skin your knee playing basketball at school. At first, everything seems fine, but after a couple of days you notice that it is getting more red and swollen. The redness starts creeping up your leg, you start having fevers and chills, and you feel dizzy and lightheaded. You go to the doctor, who diagnoses you with an infection of flesh-eating bacteria. She tells you that you need to be admitted to the hospital immediately. *By the time you get there, you are in shock from the infection, will need emergency surgery to fillet open your leg to get rid of the infected tissue and should expect to spend a considerable amount of time in the Intensive Care Unit...*

You are recovering from your broken hip, but because of the pain you have not been getting out of bed much. Your cough is worsening over the past few days, and despite trying to cough out the phlegm you are having more trouble breathing. You start to get some fevers, and the doctors diagnose you with a pneumonia and start you on antibiotics. However, despite this treatment, a few days later your breathing gets so difficult *that they need to put you on a ventilator and transfer you to the Intensive Care Unit...*

Most of us do not spend much time thinking about acute inflammation or critical illness. Maybe we should. Regardless of the disease that scares you most, or of what statistics say people die from, in this day and age the final common pathway is nearly invariant: an encounter with the health-care system, and if you are sick enough, care in an Intensive Care Unit where you will be the beneficiary of the best life-saving technology that can be provided (at least as long as you live in a developed country). The disease process that puts you there, however, nearly always stems from the same source regardless of what started it: your body's inflammatory response to some initial insult that, if it becomes disordered, can lead to the rapid, progressive failure of multiple organs. Depending on myriad factors, you could either find yourself spending a long time in the hospital, after which time you may need further convalescence. Or you could die.

This is something of an ugly little secret, swept under the rug, lost in the shuffle as doctors and biomedical researchers focus on the individual diseases that drive toward this final common pathway. So why has a highly developed society like ours not yet solved the puzzle of critical illness? *We lay the blame squarely on the current state of biomedical research.*

Biomedical research today lives in a world that, in many ways, is strikingly similar to that of the French Revolution as described by Charles Dickens. It is a time of incredible promise, resulting from unprecedented advances in technology that has led to a previously inconceivable degree of characterization of biological systems. The window into the essential components and machinery of life has never been so wide and the resulting view so sharply defined. However, as this embarrassment of riches carries with it a wealth of expectations, so too is there a corresponding chasm of disappointment when those expectations are not met. How can this increased ability to peer into the workings of biological systems be translated into actionable knowledge that can be used to aid mankind? This is the *Translational Dilemma* that faces biomedical research: the ability to effectively translate basic mechanistic knowledge into clinically effective therapeutics, most apparent in attempts to understand and modulate "systems" processes/disorders, such as sepsis, cancer, and wound healing. Unfortunately, the Translational Dilemma appears to be cropping up more and more often, as, paradoxically, a greater understanding of the processes that lead to the transition from health are known, the more intractable trying to manipulate those processes seems to become. Thus, the current situation calls for a reassessment of the scientific process as an initial step toward identifying where and how the process can be augmented by technology. The US Food and Drug Administration report: "Innovation or Stagnation: Challenge and Opportunity on the Critical Path to New Medical Products" [1], clearly delineates the steadily increasing expenditure on Research and Development concurrent with a progressive decrease in delivery of medical products to market. In many ways, the biomedical community can be viewed as standing on a ledge in a canyon, able to see the upper rim above but faced with no path in that direction, and at the same time fearing the depths that lie below.

Nowhere is the Translational Dilemma more apparent than in the reductionist approaches to understanding and manipulating the acute inflammatory. Just for context, the developing world is a morass of acute and chronic infections, traumatic injuries due to lack of civilian safety infrastructure as well as military conflict, and nonhealing wounds due to multiple factors that include malnutrition and other man-made and natural causes [2,3]. We are perhaps more familiar on a daily basis with inflammation in the industrialized world. We have our share of infections, trauma, and wounds. In our case, these diseases are complicated by our profligate lifestyles, which lead to diabetes, and obesity. We live longer, but our lives are in some ways less healthy than ever before, and our long lives are fraught with aging-related diseases such as cancer, arthritis, and neurodegenerative diseases [4]. Our better (and more expensive) hospital system as compared to that of the developing world also means that many patients will spend at least some time in an intensive care unit due to organ failure that is related to, and likely at least in part driven by, maladaptive, whole-body inflammation.

In our own work, we have focused on one key aspect of inflammatory disease, namely acute inflammation following critical illness such as sepsis, trauma, and wound healing. So, we must first set briefly the stage with regard to what these diseases entail. Critical illness can result directly from trauma, hemorrhagic shock, and bacterial infection (sepsis). On its own, trauma/hemorrhage is a leading cause of death worldwide, often leading to inflammation-related late complications that include sepsis and multiple organ dysfunction syndrome/multiple organ failure (MODS/MOF) [5–7]. Sepsis alone is responsible for more than 215,000 deaths in the United States per year and an annual health-care cost of over $16 billion [8], while trauma/hemorrhage is the most common cause of death for young people in the United States, costing over $400 billion annually [9–11].

Acute inflammation plays a direct and driving role in the pathophysiology of these conditions, producing hyper-inflammation initially, and the immunoparalysis at later phases. At a basic level of understanding, there have been numerous advances in defining novel molecules, signaling and synthetic pathways, and gene regulatory networks contributing to inflammation. However, these advances were produced and remain in scientific silos that were unable to connect and integrate their accumulated knowledge, and therefore missed an essential, systems-level understanding of the inflammatory response. An unfortunate consequence of this fractured community is reflected in the dearth of available therapeutics for these deadly and costly diseases; as of the writing of this book, there is not a single approved therapeutic targeting any component of the inflammatory pathway for these diseases. This fragmentation is further reinforced by popular notions of inflammation, where it is invariably cast as a negative thing to be overcome. There is a poor recognition of the many individual-specific manifestations of inflammation, and a lack of understanding about the favorable and important roles that inflammation plays in our minute-to-minute adaptive responses to stress, injury, and infection. In short, inflammation and related phenomena are part of a complex biological/physiological/sociological system that has, to date, generally defied a unifying understanding.

It is now beyond doubt that inflammation, with its multiple manifestations at the molecular, cellular, tissue, organ, and whole-organism levels, drives outcomes, both positive and negative, following injury and infection, and can lead to diverse manifestations of chronic diseases such as rheumatoid arthritis, neurodegenerative diseases, the metabolic syndrome, and cancer. It is very important to mention the fact that inflammation is not in and of itself detrimental. Well-regulated, self-resolving inflammation is necessary for the appropriate communication and resolution of infection and trauma, and for maintenance of proper physiology and homeostasis. Though properly regulated inflammation allows for timely recognition and effective reaction to injury or infection, disorders of acute inflammation accompany trauma/hemorrhage, sepsis, the wound healing response, and many chronic degenerative processes. In these settings, inflammation of insufficient, disordered, or overabundant, and this mismatch between the underlying reason for initiating inflammation and the way that inflammation progresses can impair normal physiological functions. This paradox of a robust, evolutionarily conserved network of inflammation whose very structure may lead to disease [12] has resulted in its near ubiquitous involvement in those diseases that most dramatically manifest the Translational Dilemma. Indeed, most evidence suggests that either insufficient [13] or self-sustaining [14] inflammation drives the pathobiology of trauma/hemorrhage, sepsis, inadequate or exaggerated wound healing, and a host of disorders at the molecular, cellular, tissue, organ, and whole-organism levels. These complex interconnections generate—or manifest in, depending on your point of view—a series of nested, interacting and balanced negative and positive feedback loops (Figure 1.1.1).

It is therefore not entirely surprising that merely suppressing inflammation is an ineffective therapeutic strategy other than in extremely severe or very benign settings. As an alternative, we suggest that controlling and reprogramming inflammation may allow us to reap its benefits while minimizing its detrimental aspects. However, the paradigm under which most of science operates is reductionism, and reductionism has largely failed to provide a rational approach by which to accomplish this goal. In addition to the multiscale complexity inherent in its organizational structure, inflammation manifests very differently based on personalized factors. These factors include individual features of the initial inflammatory perturbation, the individual's demographic and disease histories (including genetic predispositions and setpoints/thresholds for inflammatory processes), and the impact of environment and clinical care.

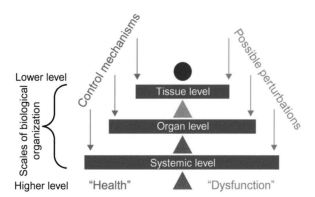

FIGURE 1.1.1 **Multiscale control structure of inflammation.** This figure demonstrates the tiered scales of biological organization. Control mechanisms (such as inflammation) attempt to balance insults/perturbations that threaten the health state (abstractly represented as the purple circle). Balance occurs at multiple tiers, and the multiscale nature of the control mechanisms allows for considerable robustness of the system to perturbations. Note that the control mechanisms themselves have a complex structure that can shift the balance as well. Source: *Reprinted from Ref. [15].*

Over a decade ago, there was recognition of the complex interplay between inflammation and physiology in critical illness and therefore also of the need to apply complex systems approaches such as computational modeling to unravel this complexity [16,17]. The use of mathematical and computational modeling of biological systems has become more common with the development of the systems biology and computational biology fields; the former generally modeling the behavior of intracellular signaling pathways and gene regulatory networks, the latter generally focusing on the development of correlation/pattern identification methods for large gene/protein data sets. We suggest that these types of analysis could be extended to decipher the multidimensional puzzle of inflammation and its consequences, and that, when geared toward practical applications, these methods hold the potential to transform the entire process of health-care delivery from preclinical studies, through clinical trial design and implementation, to personalized diagnosis and therapy, and ultimately to long-term care.

We and others have envisioned a rational, systems engineering-oriented, computationally based investigatory framework, *Translational Systems Biology*, that can integrate data derived from basic biology experiments with preclinical studies and clinical studies, and ultimately lead to the development of strategies for rational inflammation reprogramming [16,18]. Translational Systems Biology involves using dynamic mathematical modeling based on mechanistic information generated in early-stage and preclinical research to simulate higher-level behaviors at the organ and organism level, thus facilitating the translation of experimental data to the level of clinically relevant phenomena. Below, we provide the tenets and components of Translational Systems Biology. These concepts may seem a bit arcane at this point, but our presentation here sets the stage for their further explanation and discussion throughout this book.

PRIMARY GOAL: FACILITATE THE TRANSLATION OF BASIC BIOMEDICAL RESEARCH TO THE IMPLEMENTATION OF EFFECTIVE CLINICAL THERAPEUTICS

Primary Design Strategies:
1. Utilize dynamic computational modeling to capture mechanism.
2. Develop a framework that allows "useful failure" à la Popper.
3. Ensure that the framework is firmly grounded with respect to the history and philosophy of science.

Primary Methodological Strategies:
1. *Use dynamic computational modeling to accelerate the preclinical Scientific Cycle by enhancing hypothesis testing, which will improve efficiency in developing better drug candidates.* This includes the use and integration of methods applied to the other phases of the Scientific Cycle, i.e., high-throughput experimental platforms for real-world validation, generation of "omics" data sets representing enhanced observational capability, Big Data analysis to identify new correlations. Dynamic knowledge representation plays a big role here.
2. *Use simulations of clinical implementation via in silico clinical trials and personalized simulations to increase the efficiency of the terminal phase of the therapy development pipeline.* This focuses the targeted modeling goal on generating populations of simulations that can mirror how the biology manifests at the patient/clinical level, and looks to patient/epidemiological/clinical data as validation/verification metrics.
3. *Use the power of abstraction provided by dynamic computational models to identify core, conserved functions, and behaviors to bind together and bridge between different biological models and individual patients.* This will provide a formal and rationale guide to assess what aspects of mechanistic biology can be considered similar through the range of preclinical and clinical biological systems, and reduce the set of unknown factors that can be invoked to try and explain individual heterogeneity. Intrinsic to this goal is the need to dynamically model states of baseline health from whence disease states arise.

What Translational Systems Biology is not (which is not to say that the following are not laudable, or even necessary goals):
1. Translational Systems Biology is not using computational modeling to gain increasingly detailed information about biological systems.
2. Translational Systems Biology is not aiming to reproduce detail as the primary goal of modeling; level of detail included needs to be justified from a translational standpoint.
3. Translational Systems Biology is not aiming to develop the most quantitatively precise computational model of a preclinical, or subpatient level system.
4. Translational Systems Biology is not just the collection and computational analysis of extensive data sets in order to provide merely a broad and deep description of a system, even if those date sets span a wide range of scales of organization spanning the gene to socioenvironmental factors.

We emphasize that the implementation of a program of Translational Systems Biology does not preclude ongoing reductionist experimental investigations, the development and utilization of systems biology approaches to quantify fine molecular detail, or progress and utilization of Big Data-oriented computational biology work. All these approaches represent vital aspects within the Scientific Cycle, and as such have important roles to play in the current biomedical research environment. But, given our recognition that these strategies alone are not sufficient to meet the Translational Dilemma, we have crafted the description of Translational Systems Biology to limit overlap with those pursuits, and by so doing emphasize what is missing from all those approaches.

This book will introduce and demonstrate the Translational Systems Biology approach in three separate phases. The First Phase (represented by Section 2) describes the "Why?" behind the development of Translational Systems Biology. This section consists of primarily a historical, philosophical, and social survey of science, how it is used in biomedical research, and how those factors have led to the current biomedical research environment. The Second Phase (represented by Section 3) describes the "What?" in terms of how we propose to solve the issues identified at the end of the First Phase. This section introduces and provides a description of Translational Systems Biology and the primary intellectual strategies utilized in its pursuit. The Third Phase (represented by Section 4) described the "How?" in terms of the methodological approaches to implementing Translational Systems Biology. This section provides a more detailed survey of the specific methods and biological processes we have used and targeted in the development of Translational Systems Biology as applied to the study of inflammation. We will close with a discussion of how we suggest such a research program can come into being, given the sociopolitical and economic inertia present in the biomedical research community today.

Throughout this book, we will harken back to the opening quote from *A Tale of Two Cities* in relating the dual nature of much of what we will discuss; this is true not only for the sociohistorical aspects of the need for Translational Systems Biology, but also, interestingly, mirrored in inflammation itself. This means also that while we recognize that many of the barriers we need to overcome to move biomedical research forward are not placed by malice, negative products of good intentions are no less an impediment that those placed by intention. In fact, when the case is that such impediments arise out of the pursuit of what appear to be sensible reasons, it can make it that much more difficult to overcome these barriers. We hope that the reader will find us up to this challenge.

HOW TO APPROACH THIS BOOK

This book, like any book, has an intrinsically linear structure: it starts and progresses from page to page until you reach the end. But, just because a book has this particular form does not mean it has to be read in such a fashion. As the title suggests, this book provides a roadmap for how biomedical research can rise to meet it current challenges, but given the complexity of the nature of the Translational Dilemma, it also stands to reason that the solution is not a straight line. A wide array of issues are involved in the formulation of a strategy to address the Translational Dilemma, extending from the philosophical basis of science, to the procedural aspects of integrating *in silico* approaches with traditional experiment, to socio-operational issues involved with the current academic environment and the training of the next generation(s) of multidiscipline-capable scientists, as well as the economic and political incentives and disincentives that have prevented the large-scale application of the solutions we propose herein both in academia and industry. This web of history, rationale, cause, effect and solutions mean that readers of this book should feel free to jump from section to section, based on his or her interest at a particular time. Given its scope, this book incorporates sections written in a range of styles, and different sections may suit the reader's different moods at different times. Chapters in Section 2 primarily take the form of essays emphasizing historical and philosophical issues related to the state of biomedical science. These narratives are intentionally provocative, putting forth potentially controversial viewpoints and opinions in a tone and format usually not associated with scientific reports; our goal is to stimulate readers to try out alternative ways of thinking about the scientific endeavor (i.e., outside the infamous "box"), particularly as this pertains to the quest to improve human health. Some of these narrative sections include a suggested reading list for those interested in investigating the concepts raised in more depth. We indeed hope that these sections will stimulate the reader to question long-held beliefs. Chapters in Sections 3 and 4 take a form closer to traditional scientific review articles, presenting specific concepts and methods and providing examples from existing work in Translational Systems Biology as applied to the study of acute inflammation and critical illness. We consider the content of Sections 3 and 4 the objective evidence used to substantiate the claims and conclusions presented in the essays. As such, these sections contain lists of references consistent with scientific reports. To reiterate what we suggested above, readers are encouraged to take a nonlinear approach to this book, jumping around as their mood and intellectual inclination suggest. If, after reading this book, we introduce some reasonable doubt about the how's and why's of the current *status quo* of the biomedical enterprise; if we stimulate an "a-ha" moment or just cause there to be a question; if we can change the dialog just a bit; or, if our book reaffirms the reader's belief that, despite everything we have said, the system is just fine as it is; then, we have accomplished our goal.

References

[1] U.S. Food and Drug Administration. Innovation or stagnation: challenge and opportunity on the critical path to new medical products; March 2004.

[2] Brennan RJ, Nandy R. Complex humanitarian emergencies: a major global health challenge. Emerg Med (Fremantle) 2001;13(2):147–56.

[3] Hewitson J, Brink J, Zilla P. The challenge of pediatric cardiac services in the developing world. Semin Thorac Cardiovasc Surg 2002;14(4):340–5.

[4] Bittles AH, Black ML. Evolution in health and medicine Sackler colloquium: consanguinity, human evolution, and complex diseases. Proc Natl Acad Sci USA 2010;107(Suppl. 1):1779–86.

[5] de Montmollin E, Annane D. Year in review 2010: critical care—multiple organ dysfunction and sepsis. Crit Care 2011;15(6):236.

[6] Gustot T. Multiple organ failure in sepsis: prognosis and role of systemic inflammatory response. Curr Opin Crit Care 2011;17(2):153–9.

[7] An G, Namas R, Vodovotz Y. Sepsis: from pattern to mechanism and back. Crit Rev Biomed Eng 2012;40:341–51.

[8] Angus DC, Linde-Zwirble WT, Lidicker J, Clermont G, Carcillo J, Pinsky MR. Epidemiology of severe sepsis in the United States: analysis of incidence, outcome, and associated costs of care. Crit Care Med 2001;29(7):1303–10.

[9] Namas R, Ghuma A, Hermus L, Zamora R, Okonkwo DO, Billiar TR, et al. The acute inflammatory response in trauma/hemorrhage and traumatic brain injury: current state and emerging prospects. Libyan J Med 2009;4:97–103.

[10] Patton GC, Coffey C, Sawyer SM, Viner RM, Haller DM, Bose K, et al. Global patterns of mortality in young people: a systematic analysis of population health data. Lancet 2009;374(9693):881–92.

[11] World Health Organization. Young people: health risks and solutions; 2011. Fact Sheet No. 345 (Updated May 2014 http://www.who.int/mediacentre/factsheets/fs345/en/).

[12] Nathan C. Points of control in inflammation. Nature 2002;420(6917):846–52.

[13] Namas R, Ghuma A, Torres A, Polanco P, Gomez H, Barclay D, et al. An adequately robust early TNF-α response is a hallmark of survival following trauma/hemorrhage. PLoS One 2009;4(12):e8406.

[14] Neunaber C, Zeckey C, Andruszkow H, Frink M, Mommsen P, Krettek C, et al. Immunomodulation in polytrauma and polymicrobial sepsis—where do we stand? Recent Pat Inflamm Allergy Drug Discov 2011;5(1):17–25.

[15] An G, Nieman G, Vodovotz Y. Computational and systems biology in trauma and sepsis: current state and future perspectives. Int J Burns Trauma 2012;2(1):1–10. [PMID: 22928162.]

[16] An G, Vodovotz Y, editors. Complex systems and computational biology approaches to acute inflammation. New York, NY: Springer; 2013. http://dx.doi.org/10.1007/978-1-4614-8008-2.

[17] Vodovotz Y, Billiar TR. *In silico* modeling: methods and applications to trauma and sepsis. Crit Care Med 2013;41:2008–14.

[18] Vodovotz Y, Csete M, Bartels J, Chang S, An G. Translational systems biology of inflammation. PLoS Comput Biol 2008;4:1–6.

THE CURRENT LANDSCAPE: WHERE IT CAME FROM, HOW WE GOT HERE, AND WHAT IS WRONG

2.1

A Brief History of the Philosophical Basis of the Scientific Endeavor: How We Know What We Know, and How to Know More

If I have seen further it is by standing on the sholders [sic] of Giants.—**Sir Isaac Newton**

Knowledge is in the end based on acknowledgement.—**Ludwig Wittgenstein**

The game of science is, in principle, without end. He who decides one day that scientific statements do not call for any further test, and that they can be regarded as finally verified, retires from the game.—**Karl Popper**

What is science? What does it mean to think scientifically? What does it mean to "do science?" This is a book about identifying the foundations needed to answer those questions: identifying the basis for what we believe, why we believe it, and how that translates into how we behave. We propose in this book that today's biomedical science is broken, and that it has reached this critical point in part because of the trends of scientific history. Furthermore, we suggest that biomedical research also has the ability to transcend its current crisis based on principles arising out of that same history. We hope to make our point by tracing a systematic path from the foundations of scientific thought to answering questions about how we should be addressing the challenges to the biomedical research in the twenty-first century, and beyond. In our view, the rationale for Translational Systems Biology represents the results of applying the scientific process to process of science. To make our case, we must go the beginning.

What follows here is a very brief survey and explicitly noncomprehensive and selective tour through the history of the philosophical basis of science. For readers interested in a more in-depth examination of many of the individuals and concepts introduced in this chapter, at the end of this chapter we have provided a list of Suggested Additional Readings, consisting of the source materials as well as summarizing books that we have found beneficial in our own reading. To a great degree, the events and anecdotes that we use to relate our story reflect the nature of *our* story as well (we get very meta in this book). After all, every story requires some choice on the part of the storyteller about what to leave in and what to leave out. For a writer, that is just good writing; we believe that it is also essential to being a good scientist. We will see though the examples below, and in other stories throughout this book, that perhaps the greatest challenge to science is absolutism masquerading as prevailing dogma. Despite the fact that science is grounded in skepticism, all too often the prevailing scientific community of the day forgets that they, too, need to be subjected to the same scrutiny that they promote within their respective scientific disciplines. This in of itself is telling, and suggests that, at a fundamental level, the nature of humans and the societies they form need to be subjected to constant and persistent vigilance, and primed for course correction.

In the beginning, to paraphrase the Bible, there was religion. Without getting into issues about the specifics of each religion, the fact that religion is a ubiquitous component of every human society points to its role in serving a basic human need. That need is the drive to provide an explanation for the vast and varied world about us; in short, it is a desire to find some order to the world, a way to explain to ourselves why things are the way they are. It can be argued that this desire and need to *generalize* from disparate observations has been the primary outlet for humanity's intellectual and reasoning capabilities throughout history. In many ways, the concept of intelligence can be linked to the capability to formulate useful generalizations and extrapolate lessons learned from one particular situation to other, recognizably similar situations. The power and influence of the quest for order (and more specifically, substantiate, reinforce, and justify our existing concepts of order) will show its impacts, both positive and negative, throughout our story. Humans are storytellers: the manifestation of that tendency, as applied to understanding the world around us, is to generalize our knowledge by creating stories about why things are the way they are. It is important at this point to stop for a moment and recognize the implications of such basic human drives: human beings want order in their world, they seek explanations, they construct stories, and they become invested in their stories. We will return to these themes throughout this book.

Some say science has supplanted religion as the primary means by which humans understand the world and their place in it. All too often this statement is associated with an implication of an "either/or" condition related to science and religion. Unfortunately, this false dichotomy has its origins in misunderstanding what each domain represents. Rather than thinking about religion and science as competitors, it is more constructive to recognize how each of them addresses the human desire for order by appealing to different aspects of our psyche. We propose that the fundamental difference between the two is that while religion requires faith, science requires doubt. Put in another way, religion says there is a final answer beyond which no more can be known and we must just accept its existence, not only despite not being able to prove it, but perhaps *because* it cannot be proven. In contrast, science says that there are always more questions, since doubt always remains and there is always something more to be known; once the possibility of proof/disproof disappears, so, too, does the domain of science. The scientific process can then be characterized as how humans deal with addressing their doubt (at a particular point), and recognizing that addressing that doubt just raises another set of doubts. Given our list of aspects of human nature, what is the root and history of doubt? Analysis of doubt must begin with an understanding of how we know what we know, and that entails a basic understanding of the philosophical discipline of epistemology: how much can we believe in, and what we can experience through our senses and interpret with our reason?

MODELS IN THE CAVE

One of the earliest and best-known stories about the limits of our senses in interpreting the natural world involves Plato's Cave. In this parable, the perception of reality of the prisoners in the cave is limited to the shadows projected on the cave wall by various light sources behind them. We will dispense with, for the moment, the discussion about the consequences to the single prisoner who turns around, but rather use the first portion of this tale as a representation of how we interact with the world. The parable assumes that there is a reality that is "hidden" to our most basic perceptions of what is projected on the wall of the cave; enlightenment then arises as we gradually turn around. But rather than necessarily focusing on what we find when we turn around, in terms of characterizing a process we wish instead to look at *how* we turn around: what do we see when we turn around a little bit that makes us want to turn around a little more. We propose that the process of turning around and how that turning results in various changes in the shadows is analogous to the use of increasingly artificial objects, i.e., models, to shed progressively more clarity on the true nature of the objects being projected. Viewed in this fashion, this famous parable provides an early description to how we view the world through models. By changing the projections available to our senses through the use of specific artificial objects (i.e., models), these models may cast specific types of shadows, and the comparison between the different shadows can lead us to make some inferences about the source of the light. Thus, even at this early stage, we can see that the greater the number of perspectives we can generate, the more comprehensive picture we can derive.

A critical social aspect of Plato's lesson is that it is extremely difficult to achieve the insight afforded by turning around if your peers are satisfied with their current condition (i.e., they ask you "why keep turning?") Plato goes on to add that if, even in the face of this discouragement, one were able to turn around and obtain enlightenment, an attempt to return to the cave and communicate this information is likely/certainly to be welcomed with suspicion and hostility. We would argue that in order to convince the rest of the prisoners to allow additional turning, the turner would be well served to provide some tangible benefit to the group from their endeavor. Thus, the pursuit becomes something more than just an intellectual/philosophical exercise; we have added the *applied* aspect of insight and science to the equation. We would argue that this translation of insight into practice and its feedback to the general population is a critical component of the successful integration of science into its parent society. Furthermore, in order for this applied strategy to come to pass, the turner needs to recognize when he has turned enough to be able to sufficiently operate and interact in his world; this is the difference between a pure pursuer of knowledge, and those with a more practical bent. For those of us in the latter group (which, as biomedical researchers, is where we should be), we must identify circumstances in which we can say we have turned around enough: "This set of shadows is good enough for me to live my life and help my fellows." Of course, success brings it dangers as well. Perhaps the implementation of a particular insight could prove to be almost "too" beneficial for the overall group; there is belief that all that can be known has become known. This is the pathway to dogma, a manifestation of the seductive danger that we have done enough turning, become satisfied, and are unable to realize that there is more light to be seen. These are the circumstances in which we struggle within a particular paradigm to find useful solutions, not yet recognizing the overall futility of such a pursuit. It is not easy to look beyond our current success to the point of future failure, but this is, in fact, what science requires.

EARTH AT THE CENTER: A REASONABLE MISTAKE, AND AN UNREASONABLE PERPETUATION

One of the most famous examples of the danger of being wedded to a limiting paradigm is the Ptolemic Geocentric Cosmos. The story of Ptolemy's Epicycles is well known: a view of the Universe predicated upon the concept (ultimately proven wrong) that everything rotated around the Earth, i.e., geocentricism. The epicycles referred to the presumed curly-cue movement of certain astronomical bodies as they rotated around the earth; a behavior required in order to explain how some bodies, such as Mars, appeared to move backward and forward in their progression across the night sky. Rather than rehash just how wrong this model was, and the socio-religio-political forces that supported and perpetuated this concept, we would rather look at the history of this cosmology as it affected the evolution of science, particularly in terms of how the scientific community of the time responded to the increasing acquisition of data and observations about the night sky. A major, unfortunate consequence of the eventual failure of Ptolemy's Epicycles is that it has cast a negative light on its originator. Ptolemy was, in fact, a genius in his or any time, producing critically important treatises on optics, music, and geography in addition to his work on astronomy. He provided a vital link from the Greco-Roman advances in science through the flowering of Persian and Islamic thought into the European Middle Ages. In fact, his attempts to bring mathematical order to the observed world constitute a vital step in the search for nonsupernatural/metaphysical unifying explanations for real-world phenomena. Paradoxically, to a great degree, it was his acknowledged genius that proved to be a hindrance when it came to assessing his conceptual model of the world and cosmos; as the centuries passed, his concepts formed an unchallengeable "truth" for subsequent would-be explorers of the natural world. It was within this context of accepted "truth" that subsequent astronomical observations were interpreted: the data could only be correct if it corresponded to the expectations of the general structure of the Ptolemic paradigm. In fact, as the formulae of the epicycles were tweaked over the centuries to fit the existing data, the predictive capacity of this astronomical viewpoint became stronger and stronger, so much so that had significantly greater predictive power when compared to the original Copernican heliocentric view. Today, this discrepancy is usually mentioned as a preamble to discussions of the contribution of Kepler and elliptical orbits to astronomy and physics, but doing so misses important aspects of the original collision between the geocentric and heliocentric viewpoints. First was the obvious point that the Ptolemic Universe was pretty good at predicting things! This suggests that given enough data for fitting, a resulting model can become very efficient indeed in predicting future behavior. This, in fact, is a pragmatic rationale for today's Big Data science; the predictive power of a model can be independent from the model's representation of how the underlying data were generated. Therefore, sometimes prediction is good enough. However, this notion inevitably leads to the next question: when is prediction not good enough? What happens when we want our models to do more than just tell us what is going to happen, but not why? Going back to Plato's Cave, what if we want to turn around a little more? This brings up the second aspect of the transition from Ptolemic geocentricism to the Copernican viewpoint: providing the potential for expansion of a descriptive model not only in terms of the scale of data (i.e., more and more of the same metrics), but also in terms of the scope of data able to be explained. In short, the true power of the heliocentric view was an increase in the generality of the proposed explanation; it allowed for expansion beyond astronomy into more general physical processes now interpretable through the grand work of Newton and the introduction of *theory* to the scientific endeavor.

THE SCIENTIFIC METHOD OF FRANCIS BACON

Francis Bacon (1561–1626) is often credited with being the father of the Scientific Method. As first put forth in his *Novum Organon* or "New Method" (1620), he described the use of *induction* and the steady collection of evidence

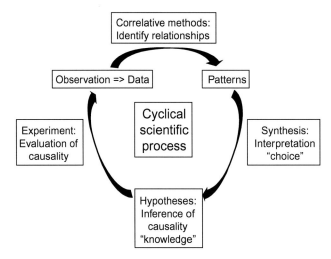

FIGURE 2.1.1   The Scientific Cycle. The iterative cycle of observation, interpretation (hypothesis generation), and experiment, originally described by Francis Bacon and the means by which scientific knowledge is generated.

as a means of establishing a description of the natural world (we will talk about induction in a little more depth shortly). Intrinsic to this process is the ranking of evidence both for and against a particular explanation, and the recognition that this is an iterative process and that some modicum of "truth" would be arrived at in a steady and incremental fashion. The steps in this process have been termed the *Scientific Method*, and the iterative nature of the implementation of this process is called the *Scientific Cycle* (Figure 2.1.1). The description of the Scientific Cycle is rightly considered to be the origin point of modern science, and we do not think it an overstatement to say that the adoption of the Scientific Cycle is responsible for the sum total of technological achievement since the seventeenth century. We will return again and again to the fundamental role of the components of the Scientific Cycle throughout this book in order to provide a foundational basis for the strategies we propose.

However, Bacon realized that, in the real world, one could not count on the perfect execution and outcome of any given process; living at the dawn of the Enlightenment, he was well aware of the frailties in the human condition. Therefore, in addition to the description of an iterative cycle of observation, categorization of evidence and refinement of hypotheses, Bacon also identified specific hindrances to the effect execution of this process, which he termed "Idols of the Mind." This list of his is striking in its timelessness:

1. Idols of the Tribe (*Idola tribus*): This refers to the human tendency to look for more order and regularity in systems than may truly exist, and leads to the tendency for people to follow their preconceived ideas about things.
2. Idols of the Cave (*Idola specus*): This refers to the tendency for an individual's personal likes, dislikes, and desired viewpoints and outcomes to affect their reasoning.
3. Idols of the Marketplace (*Idola fori*): This is due to ambiguities that arise from the use of words in science that have a different meaning than their common usage.
4. Idols of the Theater (*Idola theatri*): This is the following of academic dogma and letting those known facts preclude asking questions about the world.

Sound familiar? Herein lie the roots of *cognitive bias*; a term unfortunately often applied to those that we wish to criticize, but less frequently turned upon ourselves and our beliefs. The robustness of Bacon's overall process has been clearly demonstrated in the scientific advances over the past 500 years, as improvements in methodology, and specifically statistics, have attempted to address some of his "Idols." However, as we will see, history has shown that these idols continue to bedevil scientific communities today.

NEWTON: THE (JUSTIFIABLE) ORIGINS OF PHYSICS ENVY

Among the great thinkers of the Enlightenment it is difficult to find an individual more impactful on science than Sir Isaac Newton (1642–1727). Known primarily for his invention (concurrent with Gottfried Leibnitz) of calculus and his authorship of the *Principia*, which introduced the Laws of Motion and tied them to mathematical representations, he also pioneered investigation of optics as well as advanced analytical geometry. He also contributed to, if not drove, the societal shift to the concept of a mechanistic universe operating under Natural Law. Deism, the concept of a disengaged watchmaker god, came to dominate thought among Enlightenment thinkers, including some of the most influential of the United States' Founding Fathers, namely Benjamin Franklin and Thomas Jefferson, who perhaps not coincidentally also happened to make significant scientific contributions. The fact that Newton was also involved in occultism and eschatological Biblical interpretation gives insight into both the range of his interests as well as the socioreligious environment in which he lived. If Bacon can be considered the originator of the Scientific Method, it was Newton who, through his accomplishments, would establish the paradigm in which that Method would be applied. Here we will focus primarily on the impact of the *Principia*, which we simply state as introducing the "how" into investigation of the natural world. To summarize briefly, in the *Principia* Newton provides an explanation of *how* the observations of the heliocentric astronomy of Kepler could be generated through interactions of the physical world, and expressed that explanation in fundamental laws that could be mathematically represented, including with the newly developed calculus. This is the origin of physics. Prior to Newton, observations could be correlated together, and some order could be brought to those correlations through mathematical expression. However, there was no explicit description of how properties of the physical world could actually generate the observed correlations. By introducing *mechanism* into the scientific endeavor, Newton not only contributed to the societal shift away from supernatural explanations of the world but also provided a pathway toward the rational development of technology through design as opposed to prototyping. After Newton, one could now move beyond generating only descriptions of the physical world to being able to ask questions about how the physical world operates. Though Bacon's Scientific Method preceded Newton's career, today we invariably cast the cycle of observation, inductive hypothesis generation, and collection of substantiating evidence in the context of Newton's mechanistic world, wherein the step of induction consists both of establishing a correlative pattern within the observations as well as positing causal mechanisms that can provide means of generating those correlations.

The content of *Principia* also introduces the concept of *theory* in science. A theory is a statement of fundamental mechanistic principles that goes beyond merely explaining how a set of behaviors *could* be so; rather, a theory defines why a set of behaviors *must* be so. Theory, therefore, should be thought of a means of limiting and constraining the range of possible behaviors of a system; a theory states that, in fact, *anything is not possible*. For instance, based on Newton's Theory of Gravitation, things could not fall up from the surface of the Earth; rather, they must inevitably fall down. By providing a mathematical basis for his Laws, Newton further established the power of formalism as a means of substantiating the general application of theory. By abstracting the components of a system, the mathematical expression of a theory could be agnostic to the specific circumstances of that system, as long as the components of the real world could be reliably represented by the math. In this fashion, the application of theory suits our desire to generalize our knowledge, and the power of the abstraction allows us to apply that theory to circumstances not directly involved in the theory's development. Inherent in this process is the recognition that, given the iterative cycle, there must be some reason to believe that a potential theory can operate in this fashion. It is from here that the standard of *prediction* as a means of establishing trust arises.

The fact that his Theory was applicable at so basic a level, i.e., the *physical* nature of things, along with the unquestioned elegance of its relationship to calculus, lent credence to the impression that Newton had achieved an explanation of the Natural World as its most basic and fundamental level. The inevitable consequence of this impression is that a search for "truth" in the physical world would invariably lead to a question of physics, and that physics, as a discipline, represented the ideal paradigm for meshing real-world observation with the explanatory power of formal representation through mathematics. This is the origin of what we term "physics envy" among other scientific disciplines; a condition that has only been reinforced through the successes of General Relativity and Quantum Mechanics. There is a prevailing impression, particularly in biology, that our investigations should aspire to physics-level characterization of the systems we study. This is doubly so if the biological system we are studying is related to a disease; our lack of a cure for this disease is often related to our lack of a physics-level understanding of the underlying biological processes. Implicit in this aspiration/envy is that the goal of investigation should invariably involve a progressive reduction of our system into simpler and simpler components. This *reductionist* paradigm has its origins in Newton, and has dominated science up until the end of the twentieth century. We shall see later how, despite its unquestionable success, we now appear to have reached a boundary in the effectiveness of reductionism, and that transcending that boundary requires overcoming Bacon's Idols.

II. THE CURRENT LANDSCAPE: WHERE IT CAME FROM, HOW WE GOT HERE, AND WHAT IS WRONG

THE PROBLEM OF INDUCTION: HUME'S EMPIRICISM

The Baconian Method is foremost a method that relies on inductive reasoning, i.e., the generalization of one observed situation or condition to represent the greater set of situations. Newton now brought mechanistic causality as part of the goal of the inductive process. In its iterative form, cyclical induction is demonstrably useful. However, can induction lead to an absolute truth, or are we just aiming for some sufficient level of trust? This question was addressed by the Scottish Enlightenment philosopher David Hume (1711–1776), who focused on the problem of induction as it pertains to a wide range of human behaviors, but for our purposes will be focused on the establishment of causality. Hume divided the process of coming to a conclusion about an observed phenomenon into either a process of "reason," which we would today consider deductive logic, versus "induction," which is the generalization of one observation to a class of phenomena. Expanding upon the mechanistic Natural Law of Newton, Hume argued for an explicitly causal universe, where every observed phenomenon was generated by some causal, mechanistic event. In other words, Hume denied the role of supernatural forces as actors in the interactions observed in the real world (again, moving beyond Newton's occultist and mystical tendencies). The accumulation of knowledge about the real world can then be represented by a collection of causal mechanisms pertaining to observed phenomena. However, it became quickly apparent that an initial condition would not necessarily result in a unique outcome, i.e., multiple possible outcomes (or mechanisms) might occur. This lack of a specific outcome removed it from the domain of formal deductive proof, which provides a unique conclusion. By demonstrating that deductive logic cannot be used to derive a result merely by knowing a cause, Hume asserted that reason alone was not a means of generating new knowledge, and therefore that induction was required to establish causality. While Bacon and Newton had demonstrated the process and success of a strategy involving induction, Hume proved that such a strategy must be applied. However, now there was a problem: how reliable was induction? The Problem of Induction refers to the question as to whether an inferred individual example of a phenomenon can lead to generalized statement about the entire class of phenomenon. Hume realized that in order for induction to lead to knowledge it required two conditions:

1. The individual example must represent all cases of the specific phenomenon. The classic example of this is the observation that most swans are white, therefore, it is highly likely that all swans are white (which is incorrect, as some swans are black).
2. The conditions under which the inferred in the past will continue to hold in the future. This is the Principle of Universality. To paraphrase: that which has been true in the past will also be true in the future.

Hume demonstrated that the Problem of Induction could not be solved using deductive logic: Condition 2 failed by being subject to circular reasoning, i.e., assuming the truth of the Principle of Universality, which could not be proved independent of that assumption. Therefore, inferred statements could never be "proven" to be true; rather they could only be substantiated to some degree of confidence by the accumulation of observed empirical evidence. This concept now ties back to the Scientific Cycle of Bacon and Newton. Since the outcome of induction could not be proven, answering questions about the natural world no longer became a search for philosophical absolute truth, but rather a perpetual task of weighing evidence and finding a "best" (for now) explanation. By removing the certainty and inevitability of deductive proof from evaluation of causality, Hume offered the promise of induction-based empirical science as the pathway to knowledge through practice and endeavor as opposed to proof.

LOGIC AND ITS TRUE/FALSE PROMISE: LOGICAL POSITIVISM, GODEL AND POPPER

Hume's empiricism appeared to be borne out by the flowering of scientific knowledge seen during the late and post-Enlightenment period. Most strikingly, the success of empiricism-guided scientific inquiry manifested in applied and translational effects on society as a whole. By the end of the nineteenth century, advances in physics and chemistry found tangible expression through the steam engine, industrialization of chemical processing, advances in metallurgy, the early electrical grid, the telegraph, the early telephone, and the wireless. The world had changed from one limited by muscle power, physical senses, and the cycles of the sun to one that harnessed energy from the basic actions of the physical world, expanded what could be seen and communicated, and extended human activity to the full 24 hours of the day and across all the four seasons. It is no surprise, therefore, that optimism regarding the possibilities of technology would rule the day. From a philosophical standpoint, this optimism found expression in the school of Logical Positivism. Hume, through his approach to addressing the problem of induction and establishing causality through empiricism, is often considered the forerunner of Logical Positivism. This school of philosophy, which found its apotheosis in the first half of the twentieth century following the publication of Ludwig Wittgenstein's *Tractatus Logico-Philosophicus* in 1921, sought to provide an overarching logical framework for epistemology and put forth strict criteria for what could be considered knowledge and legitimate topics for science. They asserted that the only knowledge suitable to be discussed as such was that which could be empirically verified. It posed the formal elegance of mathematics and logic as the only way to express scientific principles, and Wittgenstein's *Tractatus* laid out the characteristics of such an ideal scientific language. It placed unquantifiable concepts that reflected explanatory interpretation of data, such as mechanisms, theories and principles, as being outside the realm of science, and therefore subject to discussion only though intrinsically ambiguous and imprecise natural language. Only that which could be measured and observed, i.e., data, could be verified and considered to be true. At first, this strict criterion for what could and could not be considered "science" might seem opposed to the prevailing optimism we just noted above. It would seem counterintuitive that the Logical Positivists would want to limit the scope of what could be considered for scientific discourse. However, we would argue that the seemingly unbounded success of empirical science gave the Logical Positivists the freedom to believe that they could limit those aspects of the world subject to formal characterization. The success of empiricism in the nineteenth and early twentieth century gave them the confidence that even after the implementation of such stringent criteria that science would find a way. This optimism concurrently manifested in the German mathematician David Hilbert's (1882–1943) attempt to provide a foundational description of mathematics through logic.

Alas, it was not meant to be. The Incompleteness Theorems of Kurt Godel (1906–1978) proved that Hilbert's overall goal of unifying all mathematics was impossible, and ever since then investigations in mathematics, logic and, by extension computer science, have been divided into those aspects of those disciplines that can be completely knowable versus those that cannot. Unfortunately, the work of Godel and others also demonstrated that the vast majority of conditions fell into the category of those things that cannot be completely knowable. In a somewhat similar fashion, the philosopher Karl Popper (1902–1994) undermined the agenda of the Logical Positivists by demonstrating the logical inconsistency of their program. By returning to the Problem of Induction, Popper proved that, despite the operational effectiveness of empiricism in individual cases, by setting the goal of science as the *verification* of inferred relationships of empirical observations, this could not be done without making the circular argument that the Principle of Universality was true. Also, he pointed out that the Logical Positivist approach could not yield uniquely true statements, as it was subjected to the fallacy of affirming the consequent: if p leads to q, and q is true, then p is true. As such, you could never completely verify that something was actually and perpetually true. Alternatively, Popper suggested that the goal of empirical investigation should be to deny the consequent, i.e., try to prove that a particular statement was false, which was now a logically tractable task (through *molus tollens*). By putting forth this reorientation in the goal of science, Popper shifted the frame of skepticism to the overall goal of the inductive process, seeking to falsify false statements rather than to try to justify potentially correct ones. This actually returned to the root of science in terms of employing doubt. Rather than trying to abolish doubt, Popper suggested that we can only confirm it, and move on. We consider this a "meta" application of the Scientific Method, where the Method is applied in a self-reflective, recursive fashion to analyze itself.

CHARLES PEIRCE SUGGESTS TAKING A GUESS: THE ABDUCTIVE APPROACH

From the onset, with Bacon's description of the Scientific Method, the process of inference and induction has been a primary source of epistemological investigation and discussion. Intrinsic to Bacon's inductive reasoning is that it incorporates both the observation and its interpretation as a *hypothesis*. As we have seen, the Newton's introduction of mechanism into the process of induction led to a prevailing conflation of hypothesis with mechanism. The American Pragmatist school of philosophy, and in particular one of its earliest proponents, Charles Sanders Peirce (1839–1914), further decomposed the process of induction and hypothesis generation to ask "where do our explanations/hypotheses come from?" Peirce realized that the step of hypothesis formation was not performed in a historical vacuum. Rather, inferences that lead to a putative hypothesis involved the contextualization of the question within a framework of existing knowledge. In short, when faced with a new observation, the scientist would draw upon a previously known list of rules/mechanisms/relationships in order to explain the current topic of investigation. This provided, if not a specific answer, but at least a formal process to address the fallacy of Affirming the Consequent introduced earlier in discussing the Logical Positivists: that given that p leads to q, and if q is true, then p cannot be said to be uniquely true. *Abductive* reasoning removes the mandatory uniqueness criteria for this logical statement: it now states that, given prior knowledge, p can be abducted as a possible means of producing

q, where the range of possible p's is limited by prior knowledge and the researcher's experience. Peirce called this process, scientifically, "guessing." From a pragmatic and operational standpoint, abductive reasoning provides a means of expanding knowledge from one known context to one that is deemed "similar."

This type of logic does have a formal syntax and set of operations, but we are not concerned with these here; rather, we point to the importance of abduction in terms of how deeply ingrained it is (even if it is rarely explicitly recognized as so) into the current scientific process. In some ways, it is much like the "crazy uncle in the attic": the scientific community understands that guessing is a necessary aspect to the scientific endeavor, but, on its face, somehow does not seem "rigorous" enough to constitute science; i.e., we scientists do not want to think about how much we do is reliant on guessing. However, as we have noted above in terms of roadblocks to advancing science, we need to deal with what we really do and not just want we would like to think we do, and this requires us to be honest about the intrinsic components of what we do. Of course, it is not all bad news, because accepting abductive reasoning means that we continue to rely on experiment and the accumulation of evidence, but now with the essential realization that the groundwork of assumptions we operate upon is a shifting and mutable landscape.

THE MAPPING PROBLEM: BACK TO PLATO?

The introduction of Abduction and the concept of Popperian falsification reinforce the importance of *experiment*. Experiment is not explicitly noted in the Bacon's original description of the Scientific Method, but quickly became an accepted part of the iterative cycle. The separation of the inductive process into its component parts makes experiment an explicit component of the Baconian induction. We will leave discussion about what constitutes a "good experiment" to later, and will focus now on what actually constitutes an experimental science (as opposed to merely observational science; more on this later). We assert that an experiment uses a model that is a cartoon of the real-world system the experiment is attempting to study; such a model represents an abstraction in which features deemed important are emphasized at the expense of realistic detail. Thus, by definition, an experimental model is "less" than the real thing; moreover, the very fact of its usefulness is to a great degree dependent upon how it is specifically "less" than the real thing. We would argue that any model sufficiently useful to answer specific questions about a system represents a model incapable of representing all the important properties of the system. The relationship between the model and the real thing is called a *mapping*. In mathematical terms, a *map* is an operation linking members of one set to another set. Here, the vernacular noun "map" provides an ideal example of the concept. A physical two-dimensional map, be it a road map, a topological map, or a population density map, represents the visualization of data about the real world in some reduced form (even beyond the compression of three dimensions to two) that emphasizes the data type chosen. A description of the mapping relationship between a model and its real-world referent is then a necessary piece of information in being able to contextualize the information obtained from the experiment. To follow Peirce, knowing the mapping of the model allows one to identify the appropriate prior knowledge that can be used to support the "guess." It should also be evident that not only can models be mapped to reality, but they can (and should) also be mapped to each other.

Now, recognizing that we obtain much of our information about the world around us through inferences made via our experiences with multiple maps/abstractions of the real world, we find ourselves back in Plato's Cave. The shadows on the wall are projections of reality made by various reduced objects (Plato's "artificial objects"), but now, armed with the fore knowledge of their limitations, rather than obscuring the reality behind us, they can provide more insight into how we turn around, and informing us that we are turning in the correct direction. The shadows can serve as guides, and provide tangibly useful knowledge in the process. But despite their usefulness, we must remember that they are themselves potential distractions from what we are ultimately trying to understand.

So, what have we hoped to gain by taking this little tour? Each one of the stories above presents an important and relevant lesson that will run as recurrent themes throughout this book. Plato tells us that we perceive the world through shadows of reality, i.e., models. He also warns us that those bringing insight might not be received with all too favorable enthusiasm. The lessons of Ptolemic Geocentrism are that even genius needs be challenged to overcome both the inertia of prior belief as well as the seductive allure of predictive correlation. Bacon not only brought us the key description of our process, the Scientific Method, but also recognized that there existed "Idols" that perpetually need to be torn down. Newton demonstrated the power of mechanism and abstract representation, and irrevocably integrated both of those processes into the Scientific Cycle. He also paved the way for the establishment of reductionism as the existing paradigm for scientific investigation. Hume closed the door on ontological truth with his skepticism and limits on induction, but opened a pathway for science that is reliant upon empirical testing. The Logical Positivists showed what can happen if you push too far to what appears to be a logical conclusion at the cost of forgetting the foundations of your belief, something Popper pointed out as he pulled us back to a tractable viewpoint that could actually be implemented. Peirce introduced a third logical way in addition to deduction and induction, namely the abductive process that underpins the practice of science today. Together, Peirce and Popper offer two tangible operating principles for the practice of science: (i) educated guesses supported by experiments and (ii) a goal of viewing experiments as falsifying endeavors. These two principles can be unified into an overarching strategy to point a way forward that rests on firm philosophical foundations.

Having provided, through this brief overview, the basis for how science should be performed, we now turn in the next section on the very special case of biology.

Suggested Additional Readings

Source Material (In order of presentation in the chapter):
Plato, *The Republic*
Francis Bacon, *Novum Organum, or The New Organum*
Isaac Newton, *The Principia: Mathematical Principles of Natural Philosophy*
David Hume, *An Enquiry Concerning Human Understanding*
David Hume, *A Treatise of Human Nature*
Ludwig Wittgenstein, *Tractatus Logico-Philosophicus*
Kurt Godel, *On Formally Undecidable Propositions in Principia Mathematica and Related Systems*
Karl Popper, *Conjectures and Refutations: The Growth of Scientific Knowledge*
Karl Popper, *The Logic of Scientific Discovery*
Ludwig Wittgenstein, *Philosophical Investigations*
Charles Peirce, *Reasoning and the Logic of Things: The Cambridge Conferences Lectures of 1898*

Popular Books about the Source Material

As we noted at the beginning of this chapter, our survey of the history and philosophy of science is admittedly highly selective and provided with the explicit purpose of setting the intellectual stage for our description of Translational Systems Biology. As such, our survey is not intended to be a comprehensive report on the philosophy of science and its development: there are several excellent volumes that can provide the reader that information. We list below several such books that fit that description, as well as several other books that focus on specific aspects of the survey we have provided.

Wittgenstein's Poker: The Story of a Ten-Minute Argument Between Two Great Philosophers by *David Edmonds* and *John Eidinow*
Beyond Wittgenstein's Poker: New Light on Popper and Wittgenstein by *Peter Munz*
Gödel's Theorem: An Incomplete Guide to Its Use and Abuse by *Torkel Franzén*
Gödel, Escher and Bach: An Eternal Golden Braid by Douglas Hofstater
Masterpieces of World Philosophy by Frank MacGill
The Discoverers by Frank Boorstin
From Dawn to Decadence: 500 Years of Western Cultural Life 1500 to the Present by Jacques Barzun

II. THE CURRENT LANDSCAPE: WHERE IT CAME FROM, HOW WE GOT HERE, AND WHAT IS WRONG

2.2

A Brief History of Biomedical Research up to the Molecular Biology Revolution

Animals are classified as follows:
  1. *those that belong to the Emperor,*
  2. *embalmed ones,*
  3. *those that are trained,*
  4. *suckling pigs,*
  5. *mermaids,*
  6. *fabulous ones,*
  7. *stray dogs,*
  8. *those included in the present classification,*
  9. *those that tremble as if they were mad,*
  10. *innumerable ones,*
  11. *those drawn with a very fine camelhair brush,*
  12. *others,*
  13. *those that have just broken a flower vase,*
  14. *those that from a long way off look like flies.*

*Celestial Emporium of Benevolent Knowledge—**Jorge Luis Borges**' fictional taxonomy of animals from his 1942 short story The Analytical Language of John Wilkins.*

Cast in the overall context of scientific endeavors, biology occupies contradictory positions as perhaps the most basic, most investigated, and yet least rigorously developed discipline. One of the most basic observations of the world is that there are things that are alive and that there are things that are not. The fact that this seemingly sharp dividing line for characterizing the world has been a constantly moving and increasingly nebulous line further complicates the placement of biology in relation to the physical sciences of physics and chemistry.

Biology is the study of living things. Encapsulated in that simple definition are two overarching processes that influence biological study even in the present day: observation and categorization. First, one observes that there are things that appear to be alive. It then immediately follows that you start making a list of the observed characteristics that cause a thing to be either living or not. You then observe that there are differences among the living things, that some groups are more similar to each other, while others are considerably different. Lists of characteristics are made (sometimes not substantially less fanciful than the one that precedes this chapter), categories are created and defined, and organisms are placed into these categories. So basic is this process of categorization that, in Genesis, God specifically tasks Adam with naming the animals, making him the first biologist listed in the Bible. From the beginning of biology, nearly all discourse has involved discussion, disagreement, and debate about the composition of these lists of characteristics, the resulting categories, and the placement of individual types. Interestingly enough, the fact that defining the most basic characteristic of biology, i.e., "What is alive?" remains an open topic of inquiry points to the practical challenges of this task. Controversy at even this most fundamental level reinforces the notion that the permutations of categorization, even if proceeding in a more methodical fashion than Borges' taxonomy, are endless. In fact, perhaps the main reason Borges' list, with its seemingly random construction, strikes us as being so odd is it violates the basic human urge to find order in the world.

Having touched upon the manifestation of this desire for order as a driving force in our discussion about the history of science in general, we will now examine the role of categorization in the development of biology as a science. As with our survey of the philosophy of science, we adopt a highly selective review of some of the principle points in the history of biology, with the specific goal of trying to see the origins of the hallmarks of the current biomedical research environment.

The study of living things can be seen to have three distinct threads. Clearly there is considerable crossover among these threads, but as we will see, they actually represent distinct disciplinary paths. The first, which we have touched upon above, is the task of categorization, which can generally be considered as the creation of taxonomies. The second task involves investigating how living things function; this field can generally be labeled as physiology. The third task attempts to define the source of living things: this is the study of origins and evolution.

The intersection between biology and medicine requires reconciling these three threads. Medicine as a discipline has, for much of its existence, focused on the second thread: the characterization of function. This makes perfect sense, since the goal of medicine is to restore a person to their normal function and is therefore inherently linked to physiology. Because function implies mechanisms, the study of physiology necessarily involves understanding dynamics; this casts the goal and study of biology clearly in the Newtonian concept of a mechanistic universe. But the investigation of mechanism cannot occur without having the categories of structure to build upon. Being part of biology, the science of medicine is dependent upon the basic categorization of biological properties and structures to provide guidance as to what should be done, but what is to be done is highly dependent upon some concept of what is wrong, i.e., identifying the dys*function*. This dual nature of medicine is still evident today in the two general categories of subjects present in the medical curriculum: anatomy (concerning description of structures) and physiology (concerning description of function). We suggest that there is a cyclical relationship between these two tasks, in which one perspective represents the cutting edge of progress for a period of time, and then the situation is reversed. We further propose that this cycle takes place at different levels of organization that mirror the prevailing general biological research of the day, initially starting at the organism level, and then becoming more basic with advances in technology. As such, the level of physiology one can describe is tied to the current state of structural categorization; thus, the categorization must precede the description of function (obviously).

The need to describe function is where the importance of dynamics comes in, as dynamic characterization is inherent in the description of causal mechanisms, and characterization of mechanism is the precondition for developing interventions. We would argue that it is the need to characterize dynamics that separates the First Biological Thread of developing taxonomies from the Second Biological Thread of studying physiology. We live in an era of technologies that can provide structural and component description of biological systems at increasingly higher resolution. Paradoxically, we suggest that this abundance of riches has created an imbalance between the ability and desire to categorize versus the need to describe function. The aforementioned two historical threads must be rebalanced in order to begin to address the current challenges in biomedical research. Moreover, we will also see that the Third Biological Thread, evolution, will also need to be integrated into the biomedical research paradigm, and we will propose how this can be accomplished. Again, we look to the history of these processes to identify the historical context of the current paradigms in biological and biomedical research. Having done this, we hope to delineate the preexisting factors that hinder us from accomplishing the goal of rebalancing the relationship between structure and function.

As with our discussion on the philosophy of science, we turn first to the Greeks for insight into the origins of our Western tradition. Looking at the works of Aristotle, the Father of Natural Science, and those of Hippocrates, the Father of Western Medicine, provides insight into how early this division arose. First and foremost, Aristotle's primary lasting impact in biology is through his systematic and orderly approach to categorizing living things. He based his classification on his observations of physical attributes, not only in terms of their external manifestations, but also employed dissection to gain insight into the common forms that lay beneath the skin. In so doing, he set the precedent for looking for "deeper" structures underlying what is evident only to the naked eye. His investigations into how living organisms functioned were reflections of the structures he identified, and served primarily to help him create his categories both of organisms (for instance those with closed circulatory systems versus those with open circulatory systems) and of organs (with their imputed roles). Less admirably, Aristotle also put forth that there was a progression from simpler organisms to more complex ones, the Great Chain of Being, with more complex organisms representing a transition toward more perfect forms, a process that culminating in humans. On the positive side, this can be seen as a description of proto-evolution, and the roots of phylogenetic trees. Unfortunately, on the negative side, Aristotle's concept of "progress" underlying biological complexity continues to bedevil the understanding of evolution and its outcomes, and has often been invoked by decidedly anti-intellectual, antisocial, and antiscientific pseudomovements such as eugenics, Creationism, and Intelligent Design.

To a great degree in a concurrent and parallel fashion, Hippocrates utilized the same principles of observation and categorization to characterize human disease. Various symptoms and physical maladies were seen to occur together, lead to subsequent conditions, and sometimes respond to certain interventions. The similarity in the cycle of scientific/biological investigation and the process of diagnosis should not be surprising; both represent manifestations of the desire to acquire knowledge in a systematic fashion, and lends credence to the fundamental nature of Bacon's characterization of the Scientific Method nearly 2000 years later after the Greek pioneers. One significant difference, however, is the inherent medical goal of needing to characterize function. A diagnosis of disease is not enough; rather, there should also be a strategy for doing something about it. This points to a fundamental difference in the endpoint goals between biological and biomedical science: medical research is intrinsically applied knowledge, in which the explicit task of medicine is to attempt to improve human health. Therefore, biomedical research can be considered inherently "translational," as the goal is to translate biological knowledge into medical interventions.

Taken together, the Greek experience is interesting in the apparent early parallel developments of these two fields. Such is the effect of the Greek historical legacy that this circumstance, by and large, persisted for centuries. Certainly, the Greeks realized that humans were also biological creatures, but the embedded concept of human "specialness" (still an ongoing socioreligious issue today), and the explicitly applied nature of medical investigation, helped create a divergence between the trajectories of general biological research versus medical investigations. General biological investigation was dominated for a large portion of its history by the fields of zoology and botany, which focused on the discovery and categorization of organisms. Medical investigations (if we can call them that), at least essentially up to the nineteenth century, attempted to characterize function by operating mainly through trial and error, without a process of understanding how to acquire knowledge of fundamental principles of disease. We can see the legacy of both of these paths today in the gulf between basic science and clinical medicine.

Basic science research has followed the legacy of general biological science, with a root emphasis on categorization, identification of structure, and description. The general historical trend has been one in which biology has advanced by being able to execute those tasks at ever-finer degrees of resolution (see Aristotle's dissections, for instance). Prior to the invention of the microscope, classification of species in zoology and botany comprised the bulk of biological investigation. Expertise was characterized by the degree of detail provided in description, identifying the subtle differences that separated this species from that species. Attempts to understand the structural order underlying the diversity of life focused on detailed description of what was shared and what was different between organisms. The invention of the microscope and the discovery of the microscopic world brought with it an opportunity to describe and characterize at an increased level of resolution, with the same skills and processes for defining detailed descriptive lists now translated into the field of Histology. Thus, this emphasis on descriptive characterization can be seen as embedded historically in the practice of biology: where once naturalists counted feathers, noted patterns of scales, marked the shape of leaves, and listed the configuration of toes, today geneticists produce lists of base sequences and single nucleotide polymorphisms. The key point here is that for the vast majority of the history of biology, the emphasis was on identifying what made things unique and distinct. Thus, expertise became focused on the encyclopedic characterization of what made things different from each other, rather than emphasizing commonalities among them.

Why do we think this is significant? We believe the result of this particular historical thread is that it has caused biology to emphasize treating its subjects as special, unique cases, with the focus being placed on the differences seen across biological systems. The result of this mindset is that it has led to a conceptual paradigm that resists attempts to find unifying principles among biological systems. Is this the only influence on biological thought? No, certainly not; above we have already noted the existence of the two other historical threads of function and origins. But we assert that the primary nature in biological science of the urge to describe and categorize places a significantly impactful (though likely subconscious) bias toward what the basic science research community considers expertise in their particular field. This is exactly how Bacon's Idols of the Theater would manifest.

As we have noted above, the primacy of description and categorization is not unearned: these tasks must precede the investigation of Second Biological Thread (function) and the Third Biological Thread (origins). Certainly, both the Second and Third Threads advanced (and advanced knowledge) in the millennia following Aristotle and Hippocrates. However, the history of medical "science" prior to the nineteenth century is notable for what we today would perceive as superstition, magical thinking, and barbarism. To us, the invocation of diseases arising from disturbances in various bodily "humors" (blood, yellow bile, black bile, phlegm) themselves tied to four "universal" elements (air, fire, earth, and water, respectively) strikes us as fanciful. To design therapies based on this understanding of disease, i.e., bleeding a patient to treat a fever (heat = air + fire = blood + yellow bile), to us seem ignorantly barbaric. It certainly did not help that for a significant period of Western European history, certain diseases were seen as the physical manifestation of moral shortcomings, a symptom of the human tendency to fall back on religion if corresponding knowledge did not exist or was not accessible. However, we think it would be unfair to cast too much aspersion to these practices given our relatively privileged cultural and historical perspective. Within the context of their time, these practitioners of early (and, may we say, prescientific?) medicine were operating using their reason to the best of their ability given the intellectual context in which they lived. To a great degree, it is possible to draw a rough analogy between the state of pre-nineteenth century medicine and the era of Ptolemic Geocentrism: both traditions heavily relied upon a primary authority drawn from a perceived Golden Age of the Greeks, and interpreted and invoked observed functions and behavior within that given preexisting mindset.

It is a reasonable assumption that the general level and capability of human intellect has remained unchanged since the origin of the species, manifest as observational capacity, creativity, curiosity, and interpretation, all controlled of course for the context of existing societal knowledge. Therefore, we should be aware that future societies, armed with a greater knowledge base, could just as easily look back upon our time as a primitive era. (For a whimsical example, in the movie "Star Trek IV: The Voyage Home," when Dr. McCoy is engaged as part of the operation to rescue Chekov from the "barbarities of twentieth century Medicine," he tosses off several *bon mots* reflecting his disdain for what we today consider state of the art.) With this in mind, we should turn our questioning minds as to how we, as a community of researchers, are able to avoid the intellectual and procedural traps that kept some very smart people doing some pretty silly (in retrospect) things.

To draw our analogy between biology and astronomy a bit further, one could characterize the difficulty in transforming biological science via the insights of the Enlightenment as there being no equivalent of Newton in the biological sphere. As we have seen, Newton brought mechanism to the physical world; such a transformation did not occur at the same time in biology. This is actually the root of the "physics envy" we discussed previously: since then, we biologists have been wishing for our own Newton and an equivalent Laws of Biological Behavior (or Fundamental Laws of Biology, a program run by the U.S. Defense Advanced Research Programs Agency [1]) to match his Laws of Planetary Motion.

One by-product of this desire is the presence of intermittent crossovers between the physical sciences and biology. A notable early example of this intersection is the interest in casting electricity as a motive biological force. Research that led to the understanding that nerves and muscles acted as conductors for electrical activity struck the imaginations of the time, and found its literary expression in Mary Shelley's *Frankenstein*.

From a positive standpoint, the recognition that living systems operated on mechanical principles led to substantial advances in physiology. Many organ systems could find their analogies in mechanical systems, the heart as a pump, the lungs as bellows, the kidneys as filters. The application of the physics of those systems led to unprecedented understanding about how biology actually functioned. And, critically, this understanding of biological function become manifest in a set of medical practices that continue to dominate today, in terms of cardiopulmonary bypass, mechanical ventilation, hemodialysis, prosthetic limbs, and artificial joints. While these are all biological replacements, they for the most part are based on a physical–mechanical interpretation of the relevant biology, and are correspondingly suited to characterization using concepts and methods from the physical sciences.

But what about fundamental driving forces specific for biology, independent (at least at first pass) from the realms of physics and chemistry? Or was deconstruction down to the level of physics description necessary to find driving principles for life? A negative manifestation of this desire to find something "special" about biology was the persistence of the concept of *vitalism*. This concept harkened back to our earliest division of the world into those things that were alive and those things that were not. Since alive things could be made un-alive, and life could spring from inert matter, it was natural to assume that there was "something" that, when added or removed, could affect this change. Today (as we will see) we know enough about the physical basis of biological systems to have (hopefully) cast aside the need to invoke mysterious supranatural forces (though, for a cautionary tale, see Chapter 2.5 on Complexity). Also, parenthetically, we see some kernels of persistence of this concept of "something undefinable" being present as we have moved away from focusing on what is alive to what makes us human, i.e., the properties that define our mind.

Fortunately, there were other significant findings about fundamental properties of biological systems that represented, at that time, a distinct departure from the physical sciences, notably the concept of Genetics and the description of Evolution. The stories of the lives and careers of Gregor Mendel (1822–1884) and Charles Darwin (1809–1882) and their respective discoveries have filled volumes much thicker than this one, so we will not recapitulate either of their lives in any detail. Rather, we wish to emphasize what each of these discoveries meant within the context of biological investigation, and how each provided what can be considered a novel approach toward examining biological systems. Both Genetics and Evolution represented the application of a process that could generate observed phenomena known for centuries: breeding of animals and plants for specific characteristics extended back to the origins of agriculture and domestication, and zoologists and botanists knew that there were clear patterns in the categories of organisms they had cataloged. The introduction of what could be considered fundamental processes that transcended specific examples within biology, genetics, and evolution began to provide biology with something akin to the Laws of Motion and the Laws of Thermodynamics. Keeping in mind the desired effect of a theory to limit the range of possibilities within the system in which the theory is applied, the effect of genetics was to place constraints on the characteristics that could be passed from one generation to the next. Similarly, the Theory of Evolution placed limits on how organisms could come to be, and how communities of organisms could be structured. It is further evidence of their impact that while these theories arose independently, they complemented each other very well.

What was still to be elucidated was a means to link the fundamental laws of biology with the processes governed by physics and chemistry. Specifically, what was the physical object that provided the function described in genetics and evolution? To answer this question, we turn to a monograph written in 1944 by the physicist Edwin Schrödinger titled *What is Life?* In this book, collected from a series of lectures given at Trinity College in Dublin, Schrödinger (he of the indeterminately alive quantum mechanical cat, quite the meme in its day) sought to address the question: "How can the events in space and time which take place within the spatial boundary of a living organism be accounted for by physics and chemistry?" To approach this question, Schrödinger synthesized prevailing thought about what the carrier of genetic information could be, and provided a logical argument linking the properties necessary for this object to carry out its function, and how such an object not only could but would necessarily arise out of the laws of physics and thermodynamics to produce life. He described that this carrier of genetic information needed to be an "aperiodic crystal" that carried the genetic information in its chemical bonds; his description of the properties of the carrier of genetic information are credited by James Watson and Francis Crick as having inspired their investigations that eventually led their discovery that DNA was such a crystal. The age of molecular biology was born.

Reference

[1] Beard J. 50 Year of Bridging the Gap: DARPA's Bio-Revolution. Defense Advanced Research Project Agency, ed.

Suggested Additional Readings

The Origin of Species by Charles Darwin
Mendel's Principles of Heredity by William Bateson
Life Acending: Ten Great Inventions of Evolution by Nick Lane
The Epic History of Biology by Anthony Serafini
The Accidental Species: Misuderstandings of Human Evolution by Henry Gee
What Is Life? by Edwin Schrödinger

2.3

Biomedical Research Since the Molecular Revolution: An Embarrassment of Riches

In that Empire, the Art of Cartography attained such Perfection that the map of a single Province occupied the entirety of a City, and the map of the Empire, the entirety of a Province. In time, those Unconscionable Maps no longer satisfied, and the Cartographers Guilds struck a Map of the Empire whose size was that of the Empire, and which coincided point for point with it. The following Generations, who were not so fond of the Study of Cartography as their Forebears had been, saw that the vast Map was Useless...—Jorge Luis Borges, "On Exactitude in science"

If the eighteenth century was dominated by chemistry and the nineteenth and early twentieth centuries by physics, then certainly the latter part of the twentieth century was the time for biology. The discovery of DNA spawned the molecular biology revolution. The seismic shift in the orientation of biology that resulted cannot be understated, and has, in fact, been expounded upon in many excellent books. Herein, we will concentrate on the both the positive and—unfortunately for the current state of biomedical research—negative aspects of this and related developments. Let us begin with the positive. For one, as Schrödinger had foreseen, the descriptive nature of biology could now be grounded in fundamental laws and processes derived from the physical sciences. The power of this approach is that biologists could finally ask "how?" The age-old goal of being able to tie description with function, all while satisfying the Newtonian goal of reductionism, for the first time seemed possible: we were now finally able to speak legitimately of understanding the "machinery of life."

In the pursuit of gaining insight into this machinery, the burgeoning field of molecular-oriented biology (which continues to dominate the general field of biology today) sought to meet two of the core aspects of the Newtonian paradigm, drawn from what was considered, subconsciously if not explicitly, the "parent" science of physics. These two aspects are: (i) the concept of fundamental laws and (ii) the manifestation of those laws as mechanisms. As we have seen in our survey of the history of science, the focus on establishing fundamental principles as seen in the Newtonian paradigm has its roots in the basic human urge of finding order in the universe. The position of physics as the most fundamental science is predicated upon the fact that physical processes underlie all of the phenomena observed in the material world. The laws of physics provide constraints on what is possible, and they percolate up into the types of chemical reactions that are allowed, and what form and output those reactions can take in various circumstances. Inherent in this mindset is that all material phenomena can somehow be *reduced* into physical processes, which in turn can be characterized by the laws of physics. This is the root of the *reductionist* paradigm, which presupposes that the pathway to understanding can come from the understanding of the underlying processes at increasingly detailed levels of resolution, perhaps all the way down to the domain of particle physics. Even if the end point of the investigation does not progress all the way down to the subatomic world, the primary exploratory process consists of studying a system by breaking it down into its constituent components, studying them, and then reconstructing the aggregate behavior. We will see below how the deconstruction process is integral to the execution of the Scientific Method but will also later show at what point this strategy breaks down in the reconstruction phase. It is this latter aspect of the molecular biology revolution to which we ascribe a fair bit of the blame for the current morass in which biomedical research finds itself.

27

Another inherent concept of fundamental laws is that they are shared across the entire range of phenomena to which they are applied. Here, biology runs into a bit of a paradox, based on its historical emphasis on characterizing differences between organisms: we know that organisms A and B are different, but they are also quite similar in some ways. The question, then, is in what way are they similar? Given the historical emphasis in biology on categorization via morphology, similarity has generally been characterized in terms of structural characteristics and by placement within the phylogenetic tree. For example, using humans as a reference point, primates are more similar to humans than rodents, rodents are more similar to humans than fruit flies, fruit flies are more similar to humans than nematodes, nematodes are more similar to humans than yeast, yeast are more similar to humans than bacteria, and bacteria are more similar to humans than very small stones. The ability to assess and potentially defend similarity becomes critical with the development and growing emphasis on experimental biology, which is predicated upon the use of biological proxies to gain insight into potentially shared mechanisms and characteristics. It is at this point that we need to introduce the concept of using models as part of the scientific process.

Central in the pursuit of mechanistic understanding is the use of the experimental method, which we have seen as initially promoted by Bacon: is only through the use of the Scientific Method that causal hypotheses (= mechanistic knowledge) can be derived. Since the hypotheses generated in the execution of the Scientific Method are the product of induction, the empiricism of Hume tells us that these hypotheses cannot ever be proven definitively, but rather only can aspire to become trusted by the collection of a "definitive" degree of evidence. The task of establishing the criteria for what could be considered "a definitive degree" became the province of statistics; we discuss statistics and statistical/data-driven modeling in Chapters 4.1 and 4.2. What at first may have seemed a relatively straightforward question: "Does this outcome/phenomenon happen more often than would be expected from random events?" has since morphed into a mathematical discipline in of itself. The power of statistics to identify correlations is undeniable, but all too often it is forgotten that the output of any statistical model can only identify an association or correlation pattern among data sets: the *reason* for that pattern being present requires the additional step of performing abduction, à la Pierce. In short, the outcome of a statistical test cannot tell you a mechanism, it can only allow you to impute or abduce one, which must then be subjected to future testing. To say "correlation is not causality" is a scientific truism, but unfortunately, like many truisms, has become so pervasive that its fundamental importance can become lost as newer, more complex and more enticing technologies arise. We will see how this is the case when we talk about the limitations of Big Data and data-driven modeling.

In terms of its impact on the development of experimental biology, the important role of statistics in determining what could be considered a "significant" experimental result necessarily fed back as constraints on the design of the experiment; i.e., the design parameters of an experiment became focused on whether it could be a statistically significant experiment. This introduces the concept of a "good" experiment. The general scientific community has a strong consensus as to what constitutes a good experiment: controlled, predetermined conditions that limit unintended variability within the experiment, a control group representing the base condition, a single perturbation/variable that separates the control group from the perturbed group, and enough samples of each experimental group such that a predetermined degree of statistical significance can be obtained. These are the minimal criteria to create a valid, interpretable experiment. Note, however, that the relationship between the experimental platform and the intended target of investigation is *not* part of this characterization. Rather, the emphasis on what constitutes a "good experiment" is only concerned with the operations of the experiment as an endeavor in of itself, not necessarily on what that experiment means in relation to the subject being studied. In fact, this type of discussion is all too often considered "speculative" and somehow nonscientific (strange, given the driving impetus for science to explain phenomena). The emphasis is placed overwhelmingly on the quality and interpretability of specific experiment and its specific results. On one hand, this is completely understandable and necessary: if the "atomic" process in the Scientific Method is the experiment, it is certainly critically important to know that the first-order conclusions drawn from that experiment are as correct as possible. However, our goal is to reemphasize that the experiment is ostensibly being performed with the purpose of studying something else, and so we provide a reminder that the experiment at hand is only a means to an end. The fact that we compelled to make this reminder draws from our (admittedly subjective) experiences in scientific conferences and as reviewers for papers and grants: all too often the critiques and discussions focus entirely on methodological issues related to the execution of the specific experiments at hand, rather than the big picture that the experiment is supposed to elucidate. Again, we recognize the importance of critique at this level, but the overwhelming emphasis on experimental procedure overshadows the even more basic question: "Does this experiment, even if performed perfectly, tell you what you think its telling you about the system you are trying to study?" The infrequency of this question and the associated discussion substantially restricts the utility of any conclusions drawn from the experiment and, at best, limits the impact of the experiment within the context of an overall (and by this we mean the overall research community knowledge base) research strategy. At worst, the lack of "speculation" about the higher-order conclusions and insights that can be drawn from the specific experiments carried out within a single study generates intellectual cul-de-sacs in which significant time, effort, and resources can be spent. (Note: In later chapters we will demonstrate that, given the structure and incentives of the current biomedical research environment, these cul-de-sacs are not actually an "at worst" scenario, but rather a desirable target for a sustained professional career. This produces an analogy that corresponds to achieving a nice, pastoral ideal, but unfortunately at the expense of advancing human health.)

A substantial part of the problem is that this step of synthesis and contextualization is extremely poorly formalized; there are no traditional formal processes for the synthesis of knowledge that correlate to those developed for experimental execution and interpretation, and as a result the step involving synthesis, integration, and contextualization of knowledge have largely been left to the intuition of the researcher. We will see that the approach of Translational Systems Biology is to target precisely this step of knowledge integration and synthesis within the larger Scientific Cycle.

But before we go there, we wish to point out another effect of this restricted view of what constitutes the goal of experimental biology. If we are following the Newtonian paradigm, which says that we can apply reductionism to study biology and that there are sufficient similarities to justify using proxy biological objects (i.e., models), and that there are design constraints imposed by the statistical standards of evaluating that experiment, then a great deal of emphasis must be placed on the creation of these biological proxy objects. This now becomes an engineering task: how can I engineer this biological system in order to meet the design constraints and features noted above? The nontriviality of this task is immediately evident in a survey of publications generated and careers built and sustained that focus on the development of different cell culture lines, genetically modified cells and animals, and the methods for molecular manipulation necessary to generate them. We recognize that this is an absolutely necessary endeavor: it is only through the generation of these biological objects that experimental biology can be realized. However, we also wish to point out what we perceive to be an unintended consequence of the emphasis on the creation of experimentally tractable models: the more they are engineered in order to meet the constraints required for a "good experiment," their ability to serve as actual proxies for the system actually being studied can become more and more impaired. In short, the engineering of biological models produces highly artificial objects that may not reproduce the important and significant characteristics of the system they were initially intended to help study. Again, we wish to emphasize that we realize the importance of this practice, but at the same time note that there are multiple assumptions inherent to the use of biological models that must be recognized.

More importantly, what can be done about this disconnect between reductionist experiments and the need to put those experiments in a larger context? What methods can be utilized to be able to leverage the knowledge generated by the use of these artificial systems? We will propose and describe later in this book how computational modeling can be used to accomplish this task, an approach that carries with it not a little irony, since computational models are often heavily critiqued for not mapping to the real world. Any yet, this mapping is often overlooked when dealing with clearly artificial biological models. To paraphrase G.E. Box, while all computational models *might* be wrong, all biological models are *certainly* wrong; the task is to find out how to make them usefully wrong.

There are certainly noncomputational means by which to identify the relationship between a model and its referent: this is the mapping process we had previously described in our survey of the history of science. Unfortunately, this process is not well formalized in biological and biomedical research; however, mapping is practiced on an intuitive and *ad hoc* level frequently, and sometimes in a tremendously constructive manner (several good examples from the field of inflammation can be found here: [1–3]). We have already seen the rudimentary level of mapping in our example of what organisms are more similar to humans than others; the main question is how are they more similar, and how can that be used to interpret an experiment. In addition to mapping the selected model organisms to the reference system of interest, it is also necessary to map the intended outputs/observables produced from the experiment: do they mean the same thing to the model organism as they do to the intended reference system (for sake of argument, let us say that is a human being; see the recent controversy regarding this point in the field of acute inflammation [4]). It should be noted that all this mapping is done at the level of phenotypic description: in order to be clear, we use the Merriam-Webster's definition of phenotype: "the observable properties of an organism that result from the interaction of the genotype and the environment." Note that this spans the range of observable metrics from number of limbs, to physiological vital signs, to circulating and tissue mediators. Note that this also includes cellular gene expression patterns, which represents the organism's response to environmental stimuli based on its germ line genomic structure (which represents its genotype). The key point here is to recognize that the output of most "omics" analysis represents a snap shot of the state of the organism/cell that may be dynamically changing over time. This goes back to perhaps earlier distinction, where the genotype represented the behavioral potential of an organism and the "programming" that was passed on through procreation, while phenotype represents the multiple forms that organism might take during its lifetime as generated by the intersection of its genetic potential and environmental factors. In order to interpret the output of a biological model, it is necessary to be as explicit as possible in determining what the mapping is between our phenotype of choice. For instance, we know the conversion from murine blood pressure to human blood pressure; therefore, we can say with some degree of certainty that one level in a mouse corresponds physiologically to another in a human.

Note that up to this point we have not addressed anything specifically related to technological advances; we have only discussed the fundamental aspects of what happens when you think biology can be decomposed into physics and chemistry and decide that is the way biology will be done. We mention this because the allure of molecular biology is so great, and it is so easy to become enamored with the latest method aimed at looking deeper and smaller and finer, that it is unreasonably easy to lose perspective of the inherent assumptions and prejudices associated with the endeavor itself. Let's face it: we love our methods. Humans are tool makers at their core, and the amount of inventiveness and intellectual capital expended at making a better mouse trap (or imaging technology, or cell line, or genetically modified animal, or network inference algorithm, or modeling language, *ad infinitum…*) makes us want to justify that task. Coupled with a reductionist paradigm, the drive for method development can take a codependent life of its own, thereby limiting the beneficial aspects of both method-making creativity and the power of reductionist science.

One need not look any further than the history of the study of DNA. It took some time after the identification of DNA as the carrier of genetic information for experimental biology engineering to develop the necessary tools to study it. First of all, the recognition that DNA was what carried the genetic code identified biology's reductionist target: as physics had the atom, biology has the gene. Second, just as physics had developed its particle accelerators in order to take the atom apart into its constituent components, so too would biology need to develop technologies to take apart the gene into the pieces and structure of DNA in order to gain insight into how it all worked. Moreover, this pursuit for more and more detailed characterization was bolstered by what was perceived as a fundamental concept (and so semi-appropriately titled) "The Central Dogma of Molecular Biology," first stated by Francis Crick in 1958. The Central Dogma can be summarized as: Genetic information flows from DNA→RNA→Protein→Function. While the limitations of the Central Dogma are now recognized, it was the primary conceptual model that drove research in molecular biology, overwhelmingly for its first half century, and still significantly in the years since. The rationale for the investigatory process developed under the directive of the Central Dogma can be described thusly: since DNA sequences determine function, if we can find out all the DNA sequences, then we can characterize all function. Implicit to this task was the rationale that if the ATCG "code" could be "broken," then the "secrets of life" would be unveiled. Experimental method engineering concerning DNA therefore focused on developing methods of finding the particular sequences of DNA within genes. Suffice it to say that the successes in developing means of base location, identification, recombinant DNA technologies, and the automation of those technologies laid the groundwork for what was thought of at the time as the Holy Grail of biomedical research, the Human Genome Project. It was believed that if we could completely map the human genome then it would provide the basis for answering the puzzles of human biology and disease. Again, the code-breaking analogy was heavily in play, with the supposition that once the cipher key was found then the messages within the code would be unveiled.

While this work was taking place, other researchers generated sets of experimental tools to characterize the molecular pathways that defined biology. These tools took the form of the artificial biological objects we noted above, cell lines, modified animals and the interventions used to perturb them, as well as the assays needed to look at the presence and levels of the molecules and pathways that drove their behavior. These investigations necessarily took the reductionist approach and followed the tenets of good experiments: single pathways were isolated, individual molecules within those pathways were identified, and their responses to different manipulations were evaluated. This type of characterization was a godsend to the pharmacology community, as it provided a reproducible process that could be used, at large scale, to engineer potential manipulations to various conditions. The preclinical research pipeline became essentially standardized, with a defined set of experimental models that through which a particular compound would be tested up until the point of conducting a clinical trial. There were incremental advances as new methods came online, but the overall structure of this research and development process remains essentially unchanged even today (which, unfortunately, is part of the problem) [5]. The unlocking of the Human Genome was sure to provide the final piece of information that would make the remainder of the drug development endeavor just a matter of being able to do enough experiments in order to solve a particular problem: finding cures would become something limited by resources, not knowledge. Here, we see the convergence of the Newtonian promise of molecular biology. We have our fundamental law: the Central Dogma. We have our process in which to characterize a disease process: reductionist experiments. We have the manifestation of what based on the Central Dogma should define a human being: the Human Genome. What could possibly go wrong?

It goes without saying that, of course, we have the benefit of retrospection as we list what went wrong. The inevitable advantage of hindsight is that we know what happened; we have information that very smart people in the past did not have. We have already seen this phenomenon when we discussed biomedical research in its prescientific phase: knowing how it comes out can make anyone *seem* like a genius or a critic (or both). So "yes," the issues we will note below now seem self-evident and totally obvious to us today. But our goal is to suggest that knowledge of the past, particularly the fundamental mental processes at play, can provide us with some strategy of how to avoid similar issues today. And, we are not the first to note the potential difficulty, if not futility, of the endeavor to gain biologically meaningful knowledge for the sum total of "omic" data [6,7].

Once the technology to sequence DNA reached a point where we were collecting more and more sequences, it was becoming apparent that the Central Dogma seemed to have some cracks. Much of the genome, in fact most of it, did not appear to actually code for proteins. Initially as a result of the assumption of the Central Dogma, this DNA was initially assumed to be "nonfunctional." The subsequent realization that these noncoding regions performed regulatory functions helped point out the limitations of the Central Dogma. Additionally, there appeared to be multiple types of RNA that, while they did not get translated into protein, also had significant function. Furthermore, many, if not most, proteins required posttranslational modifications in order to carry out their biological activity. Finally, it was also apparent that genes could have their function permanently modified without affecting their actual sequence via epigenetic changes. What had previously been thought as the only way to generate biological function, i.e., through the transcription of DNA to RNA and then translation of RNA to protein, now was just one of many possible means of converting the genetic code into biological activity. This realization added several additional layers of complexity into the interpretation of the human genetic code. Rather than breaking a code with a cipher, it now became a task akin to translating a work of prose or poetry from one language into another.

The bigger problem was with the reductionist paradigm. The overall complexity of the pathway structure in biological systems was well recognized, and there was a pervasive belief that the soundness of experimental process would be able to parse out significant effects from nonsignificant ones. To a certain degree, biomedical research counted on their reference phenotypes to coalesce the multiple interactions present into statistically significant conclusions: to make an analogy, it did not matter what was going on under the hood if the car still got you where you wanted to go. This was fine, as long as the preclinical research translated to clinical efficacy. But, unfortunately, starting in the late 1980s, the success in translating what were considered to be near certainly effective candidates started to fall off, and in dramatic ways. Potential drugs directed at sepsis, at cancer, at immunological diseases, at cardiovascular diseases, started not to perform as expected in clinical trials. In the area of sepsis (our area of expertise), the failures of a series of molecular interventions directed against what were perceived as principle mechanistic drivers of the sepsis syndrome led to what could be considered an existential crisis within the sepsis research community. These drugs should have worked, by all accepted metrics of preclinical efficacy; yet they did not. This is the classic manifestation of the Translational Dilemma: the inability to translate basic mechanistic biomedical knowledge into an effective therapeutic.

The mainstream research community cycled back to their basic strategies and focused on two main issues. The first was that they had insufficient information: the characterization of the target disease process was too nonspecific in terms of biological mechanism. The "disease" previously thought to be a single entity in actuality represented a multiplicity of potentially nonequivalent processes; therefore, understanding what to target with which drug required finer granularity in the definition of the disease. This realization led first to biomarker panels, then to DNA microarray (transcriptomic) analysis, and now various forms of "omics" characterization. Where previously a disease might be, at most, defined by a few alterations in biomarkers, now thousands of data points are used to characterize the disease "phenotype." This is the pathway to Big Data.

The second issue was that perhaps the biological proxies we used in the preclinical investigations did not actually represent what we thought they did in terms of being able to reproduce the essential aspects of the target disease process [4]. This has proven to be a more difficult challenge to address, since the fundamental limitations in terms of the design parameters for experimental platforms remain, i.e., reproducibility, reductionism, and strength of experimental signal. Here also the adoption of finer-grained phenotype description has been proposed as a means of trying to link experimental models to each other and to human conditions, though this strategy is fraught with its own challenges. While there may be some similarities in terms of organ level or systemic phenotype, and certain shared properties and functions between relatively analogous pathway modules, there are going to be specific differences between species in terms of their absolute configurations.

An example of this is the recent controversy over the mapping between the mouse and human response to trauma and sepsis [4,8]. On one hand, of course a mouse is not a man, therefore, from a detailed descriptive level one, could

II. THE CURRENT LANDSCAPE: WHERE IT CAME FROM, HOW WE GOT HERE, AND WHAT IS WRONG

not expect to directly translate the alterations in one species to the alterations in the other. At some fundamental level, however, there must be some way to identify what aspects of the responses are similar, and can therefore allow them to be compared. What is pretty certain by now, however, is that constructing a detailed parts list is not the way to do it.

Unfortunately, we see all these attempts to address the Translational Dilemma as having their roots in the tradition of biology to emphasize description. The heterogeneity of biological systems is a given; the tradition of biology suggests that greater insight can be obtained by describing that heterogeneity at ever greater detail. This is the paradigm that is sweeping the biomedical research community now, the manifestation of the current general societal fascination with Big Data, and seen in the emphasis on genome-wide association studies (GWAS), molecular profiling of tumors and patients, transcriptomics, metabolomics, metagenomics, metatranscriptomics, and whatever new "omics" becomes the latest trend. The promise of Big Data is that this methodology will provide a means of untangling the multiplicity of effects present in biology and allow an answer to emerge.

We suggest caution. We will discuss the issues related to and a place for Big Data in a later chapter, but suffice it to say, that given our penchant to look at fundamentals, we note that it should be immediately evident that the correlations offered by Big Data are only one component in the application of the Scientific Method to the Translational Dilemma. Remember our truism: correlation is not causality, and the goal of intervention requires an understanding of mechanistic causality. The step to move beyond Big Data is recognizing that we must vastly expand how we do the experiments we need to do in order to evaluate mechanism. Figure 2.3.1 depicts how the Scientific Cycle appears now in a high-throughput, Big Data environment (compare with Figure 2.1.1); there is a clear bottleneck in the iterative process at the point that requires hypotheses to be tested. The need to address the scientific step of hypothesis testing and evaluation arises out of our root-cause analysis and diagnostic approach to the Translational Dilemma, but the persistent and future consequences of this imbalance should be readily evident in this time and now. We should not have to wait until what should be recognized as an unrealistic promise of an approach (i.e., Big Data supplanting the Scientific Cycle) has been demonstrated to be so; we have the opportunity to act to implement a research strategy now that we can anticipate we will need in the future. We recognize, however, that there are multiple barriers to acting in such a manner, but we hope that the information in this book will move us away from functioning in a perpetual near-crisis mode, particularly with our health is at stake.

We've been introduced to the Translational Dilemma, and to this point we've focused on how we obtain the knowledge to decide how to potentially develop new ways of treating patients and enhancing human health, and some of the issues related to that pathway. But we've also been introduced to the translational barrier presented by

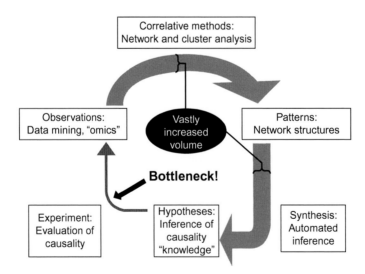

FIGURE 2.3.1 Imbalanced scientific cycle in a high data environment. The current state of the Scientific Cycle following advances in the extraction of experimental data (high-throughput "omics" methods), the development of correlative methods of data analysis (clustering algorithms, machine learning), and the generation of potential hypotheses (automatic inference). In essence, the process of acquiring data and identifying patterns in these data has become parallelized, leading to an exponential increase in the number of potential hypotheses to be tested. However, the process of experimental testing remains a time-consuming, serial process, limited in terms of resources, person-hours, and the time scale of biological processes (i.e., cells have to grow and animals need to respond to experimental stimuli). This imbalance has led to a bottleneck at the critical step of hypothesis evaluation, and currently manifests as the Translational Dilemma.

the clinical trial; now we turn our attention to examining, in a critical fashion, this last step between a good idea and an effective therapeutic.

References

[1] Nathan C. Points of control in inflammation. Nature 2002;420(6917):846–52.

[2] Nathan C, Sporn M. Cytokines in context. J Cell Biol 1991;113:981.

[3] Tracey KJ. The inflammatory reflex. Nature 2002;420(6917):853–9.

[4] Seok J, Warren HS, Cuenca AG, Mindrinos MN, Baker HV, Xu W, et al. Genomic responses in mouse models poorly mimic human inflammatory diseases. Proc Natl Acad Sci USA 2013;110(9):3507–12.

[5] An G, Bartels J, Vodovotz Y. *In silico* augmentation of the drug development pipeline: examples from the study of acute inflammation. Drug Dev Res 2011;72:1–14.

[6] Csete ME, Doyle JC. Reverse engineering of biological complexity. Science 2002;295(5560):1664–9.

[7] Mesarovic MD, Sreenath SN, Keene JD. Search for organising principles: understanding in systems biology. Syst Biol (Stevenage) 2004;1(1):19–27.

[8] Osuchowski MF, Remick DG, Lederer JA, Lang CH, Aasen AO, Aibiki M, et al. Abandon the mouse research ship? Not just yet! Shock 2014;41(6):463–75.

2.4

Randomized Clinical Trials: A Bridge Too Far?

The desired final outcome of all the research and development discussed in the prior sections is proof that the proposed intervention will actually work by improving the health outcome in a real human population taking the drug. The importance and significance of this step cannot be overstated: the evaluation of a therapy in a clinically relevant population is the final hurdle that will determine whether the drug is approved by regulatory agencies (e.g., the U.S. Food and Drug Administration), which in turn will determine whether the drug goes to production, and set into motion an intense, expensive, and influential marketing campaign to get doctors to use and/or prescribe the drug. The entire process leading up to the clinical trial is—or should be—based on the Scientific Method, and so it would seem logical to suppose that the Scientific Method would be used as the cornerstone of this final evaluation. Based on the criteria set forth in Bacon's Scientific Method, that final evaluation takes the form of a randomized, prospective clinical trial powered to determine the efficacy of the drug. Let us take a brief look at the components and justification thereof that make up what is considered the "Gold Standard" of clinical evidence.

Keeping in mind that we are still essentially operating in the Baconian/Newtonian world of evaluating a single variable (i.e., the therapy), the underlying rationale for a Randomized Prospective Clinical Trial is very familiar: it is the same rationale that underlies what is traditionally considered to be a "good experiment." Therapeutic drug and devices reach their ultimate end user—the patient—via a multistep process. This process culminates in regulatory approval (e.g., the U.S. Food and Drug Administration). This process generally consists of years/decades of basic research to identify candidate therapeutic targets, followed by sequential studies to demonstrate safety and some acceptable degree of efficacy (e.g., dosage or timing that results in greatest therapeutic benefit with least harm) in both experimental animals and humans. The final step is a pivotal (Phase III) clinical trial, which is randomized (i.e., subjects that meet predecided inclusion and exclusion criteria are recruited into either a placebo or treatment arm in a random fashion) and double-blinded (i.e., neither the clinician nor the patient knows *a priori* the study arm in which the patient is enrolled) [1–5]. The enrollment into this Phase III trial is usually not individualized in any fashion beyond the set inclusion and exclusion criteria (and, of course, the withdrawal of a patient from the study if certain predecided adverse events occur). This process is considered the *sine qua non* of clinical translation, and it has indeed resulted in numerous drugs and devices available to physicians to treat diseases.

Briefly described, a Phase III trial incorporates the following features. The study population is divided into two equally characterized groups; one group serves as the control group, representing the absence of the intervention, while the other group receives the intervention. The potential for cognitive bias (having its origins in Bacon's Idols of the Mind) is recognized, and steps are taken to try to reduce their effect. These steps include *randomization*, which the placement of an individual into control or treatment groups using some random means, *blinding*, which is the process by which the investigators do not know which group a particular individual is assigned, and being *prospective*, meaning that the data are being collected from the initiation of the study forward and therefore data analysis cannot be influenced by any prior knowledge about what outcome is desired or sought. At the end of the experiment, a statistical analysis is performed to determine if any resulting difference between the two groups is greater than would be expected by random events, and if so then the intervention is deemed to be beneficial. Note that this is the simplest description for the ideal circumstance of a clinical study. We will see shortly that there are

35

logistical and methodological barriers to the execution of a Randomized Prospective Clinical Trial, but we think it behooves us to be reminded as to what the ideal condition should be, so that we can see how the practice deviates from this ideal and thereby characterize the origins and consequences of those deviations.

As with the application of the traditional Baconian/Newtonian Scientific Method in other disciplines, the implementation of the Scientific Cycle as the intellectual basis of the Randomized Prospective Clinical Trial (and its less formalized precursors) has had incredible success over the past 500 years. The success of its informal implementation in the original arguments for antisepsis (Ignaz Semmelweis), simple battlefield dressings versus cautery (Ambrose Pare), smallpox vaccination (Edward Jenner), and public health water quality resulting from an examination of cholera epidemics (John Snow), led to a desire to formalize how such evaluations could be made. With the advent of the molecular biology revolution and the ability to begin to engineer interventions, the technical aspects of a clinical trial were able to be refined to a point where the execution for a clinical trial could be depicted as a protocol. In the 1960s and 1970s, the success of this protocol, and the drug development pipeline that led up to it, were manifested in an unprecedented explosion in clinically available therapeutics, including the current list of antibiotics, antihypertensives, statins, pain relievers, sleep aids, and pretty much every drug on the market today. We should be reminded that as with most practices that become part of the *status quo*, the application of the principles of the Scientific Method to clinical medicine works in very many circumstances; after all, people will naturally gravitate to something that works. In this respect, no one would doubt of the incredible track record of this modality as a means of improving human health.

In fact, the success of Randomized Prospective Clinical Trials led to a highly significant movement still influencing how medical practice is determined and performed: Evidence-Based Medicine. We will address this movement a little later below. However, as we have noted at various steps in this book, this traditional investigatory pipeline is drying up. Just as an increased recognition of the complexity of biological processes has complicated the application of the Scientific Method, so too did the increased significance of more complicated disease processes rise to challenge the overall efficacy of Randomized Prospective Clinical Trials. We have already touched upon the impact of the increased incidence and significance of "systems" diseases in creating the Translational Dilemma; it should be noted that it was the unanticipated failure of Randomized Prospective Clinical Trials for therapies that should, by all prior experience, have worked, that brought attention to this translational gulf. In addition, the biomedical research community became acutely aware of late consequences of the interventions thought to be safe based on the results of clinical trials: unanticipated complications or differences in efficacy started to surface only after a drug was in use in the "real" general population [6]. Finally, with increasing knowledge about influence of individual genetic predispositions cast further doubt on some of the initial assumptions inherent in the design of clinical trials, with secondary consequences arising in terms of logistical and pragmatic issues. To assess these issues in a systematic fashion, let us first look at some of the recognized limitations of Randomized Prospective Clinical Trials, and then, based on our foundational strategy throughout this book, at some of the underlying cognitive and philosophical issues with the current execution of clinical investigations.

There are two general foci of critiques of clinical trials. The first targets what can be considered failures of execution: this assumes the validity of the rationale for the clinical trial. In this case, an undesirable result is thought to be generated through a failure in appropriately designing or executing some aspect of the trial. This strategy is the most common type of critique seen, and essentially mirrors similar critiques about the methods section of an experimental biology study. Common issues raised in this type of critique are:

1. The control group and the treatment group are not *matched*. This can occur either in the process of randomization into each group (i.e., it was not *truly* random), or that somehow the groups were otherwise not treated equivalently, except for the intervention being tested, during the time of the trial (i.e., the study was not truly *blinded*).
2. The study is *underpowered*. This charge implies that there was greater heterogeneity in the selected patient cohort than was originally assumed, and that due to this the noise of the outcome was too great to be overcome by the study design.
3. *Attrition* of patient participation through the stages of the trial masks the interpretability of the results. This charge usually implies that some bias is introduced simply by being able to meet the criteria necessary for participation in the trial, be it an issue related to a patient's clinical course, their tolerance of the study drug, or ability to maintain the strict protocols of the study.

As we have noted above, one can find some variant of these comments in nearly all critiques of clinical trials. We would argue that while this discussion is important, and represents an important methodological quality check on these studies, in many ways it is too narrow in scope. Because, if we have the bigger picture in view (which is actually the case: after all, we all *want* to have better drugs), critiques of individual studies must also have the ability to suggest that there is a way to make them better. After all, it is easy to critique; it is more difficult to propose solutions. So, let us take a look at how some of the critiques above might actually be addressed (all assuming baseline competency in the original study).

First, how could one actually improve matching between control and treatment groups? As with nearly all things biological, our criterion for similarity is the descriptive characterization we have noted as the legacy of biological study. One could increase the number of criteria used to characterize inclusion into the study, creating a finer-grained description of the target population. However, this process has several notable side effects:

1. The more detailed characterization you have, the greater the number of assumptions of similarity that are required, i.e., what is the range of difference in any given criteria that can still allow the criteria to be considered similar?
2. As one increases the number of criteria that allow for the direct comparison of one group to the other, the overall pool of potential candidates is reduced, thereby leading to the charge of the study being underpowered.
3. It costs money, a lot of money.

How does a trial design avoid being underpowered? The only solutions to this problem are:

1. Increasing the number of eligible candidates, but at the risk of increasing patient heterogeneity and leading to unrecognized differences between the study groups, or introducing unanticipated noise.
2. Increasing overall recruitment, which costs money, a lot of money.

How do you avoid attrition through the process of a clinical trial? The only solutions to this problem are:

1. Front-loading exclusionary criteria, but this must be based on as detailed a characterization of candidate characteristics as possible. This would reduce initial variability, but at the expense of potentially ending up with an underpowered study.
2. Tracking the attrition as part of the study, though this leads to the issue of having to deal with multiple outcomes and endpoints, which clouds interpretation of the study.
3. Both of the above possibilities cost money, a lot of money.

At this point, it is hopefully clear that there are a set of fixed, pragmatic constraints on how optimally a clinical trial can actually be performed in the real world. The conclusion to this examination is not one that would surprise anyone who has actually performed clinical research: it is really messy, despite the best intentions and highly capable investigators. The investigators do the best that they can do, but the fact remains that the "ideal" study is quite possibly practically impossible. As unsettling at this may seem, as we have seen in our prior survey of the history of science, being able to recognize existing limitations is the only means to be able to adopt strategies to move forward; to try to deny that such issues exist does no one any service.

Unfortunately, it gets worse; as we noted above, there are an additional set of fundamental issues associated with using the Randomized Prospective Clinical Trial as the primary means by which therapies are judged appropriate for clinical use. In some ways, these issues are related to the methodological problems noted above, but we suggest that the following issues would be present even if all those critiques could be addressed:

1. *The study group is not reflective of the actual clinical use population.* As we have noted above, the Randomized Prospective Clinical Trial is an *experiment*, and the principles of good experimental design include narrowing down the study group as much as possible in order to eliminate confounding factors. We have seen the challenges inherent in trying to accomplish this goal in the section above, but the issue here is that the accomplishment of that goal for the purposes of the study is inversely related to the applicability of that drug to the general population. This is particularly important when assessing the impact of clinically available therapeutics on subpopulations that are underrepresented in clinical trials, i.e., women, minorities, and children [7]. There are certainly very real ethical, historical, and safety issues that contribute to the relative paucity of members of these groups participating in clinical investigations. Nonetheless, the fact remains that, as a result of this underrepresentation, the means by which drugs are deemed "safe and effective" does not include these individuals. We think most people would agree that the ethical and safety issues should not be relaxed in order to incorporate women and children, and while there are attempts to address historical prejudices that affect minorities (especially the African-American population), this issue represents a hard constraint on the utility of Randomized Prospective Clinical Trials.

2. *The implementation of the treatment within the study does not match what the actual use would be in the real world.* Returning again to the concept of a Randomized Prospective Clinical Trial as an experiment, the intervention must also be carried out in an as rigorous fashion as possible. This makes perfect sense: the goal is to determine whether a potential drug would actually work in an ideal circumstance. But, unfortunately, the desire to determine this efficacy in an ideal case moves the applicability of the trial further away from the general population. This problem is most notable in therapies that require multiple timed administrations, or specific concurrent actions, for the intervention to be optimally effective. While this degree of compliance might be enforceable within the context of a clinical trial, which has already selected for motivated individuals (by virtue of their mere participation) and has extraordinary support resources (dedicated study coordinators, nurses, and physicians), such conditions are not present for the general population. We are already seeing evidence that sometimes the best "biological" answer may not be the best "realistic" answer, since issues of cost, dosing, and timing limit the appropriate use of a potential therapeutic. To a great degree, this represents an engineering problem: some of these factors, at least in terms of dosing, can be features targeted at the outset. However, as we have seen the biomedical research and development pipeline is having enough difficulty in generating drugs that even work in the best of circumstances it is quite unrealistic to imagine that given the current research paradigm this factor is likely to be successfully addressed in a systemic fashion. Additionally, there is a potential beneficial treatment effect of just participating in the trial, where, in the interest of trying to reduce the variability between study groups, the control group obtains benefit from the increased attention and vigilance that goes along with participating in the study: this is known as the Hawthorne Effect. While ostensibly good for the patients (they get more attention), this problem further dilutes the potential efficacy of the therapy if it were to be implemented in the real world, i.e., the beneficial effect of the intervention is diluted by the general improved outcome across the board accomplished by greater attention. This last case does introduce some troubling potential conclusions (i.e., maybe things would be better if we just took better care of our patients?), but again, practical logistical resource constraints do limit how far that particular line of thinking can go.

These issues have been recognized increasingly by the biomedical research community over the past few decades, and some steps have been taken to try to address certain aspects, though mainly by adopting strategies derived from within a standard set of existing solutions.

The first attempted solution is in the development and application of more sophisticated statistical tools (see our discussion in Chapters 4.1 and 4.2). Remembering that the Scientific Cycle relies upon some assessment of whether an empirical result can be trusted, the means of establishing that trust has fallen to statistical analysis. What used to be a relatively straightforward assessment of whether the two groups were more different than random given one intervention (the simple case we started with) has now evolved into an extensive field of mathematics and computer science that attempts to parse data across multiple variables in multiple combinations in multiple dimensions. We will discuss some of these methods in more detail elsewhere, but suffice it to say at this point that the statistical tools employed to analyze complicated data are themselves complicated and arcane objects in their own right. We suggest that there is a potentially false promise of "truth" offered by these mathematical manipulations, one that leads researchers to invoke their various methods and algorithms with the same dogmatic passion once only seen in the domain of religious controversy. At one level, the statistical test is absolutely necessary: some ostensibly objective means of determining whether groups are the same or different is an absolute necessity. The difficulty arises when the same methods that are being promoted to evaluate the difference between two groups become the means of determining whether the method is valid in the first place. There is an intrinsically self-referential nature to statistics and one that is grounded on the fact that it is entirely phenomenological. With respect to statistics, to a great degree we are operating in premechanistic, pre-Newtonian universe, where there is no accounting for why data takes the form it has. Rather, the data becomes the only guide as to the pattern that arises, and this in turn is used to explain why the data arose in the first place. It is therefore not surprising that the truism exists that says, "There are three types of lies: lies, damn lies, and statistics." To a great degree, the retreat into statistics is akin to attempting to invoke a higher authority to provide guidance in lieu of our own understanding. We see this today with the increased slicing and dicing of data using various multivariate statistical techniques, all aimed at trying to find a metric that produces a "significant" result. Again, this is not to say that statistical tests are invalid, but rather that as they are currently employed, they are being used to search for something that one is looking for. Take, for example, the relatively recent advent of the "Lack of Inferiority" outcome present in many drug trials [8], a phrase that if spoken out loud belies its Orwellian roots (independent of the methodological problems that have been pointed out in these studies [9]).

The other primary strategy adopted by the biomedical research community to address the recognized limitations of the Randomized Prospective Clinical Trials, particularly in terms of patient heterogeneity and their application to underrepresented groups, similarly falls back on historically derived behaviors. Here, we are referring to the increasing use of genotyping and Genome-Wide Association Studies (GWAS) to attempt to both obtain finer granularity in the description of individuals (which, as we have seen, is biology's default strategy), as well as potentially making inferences of subgroup similarity across a wider population. As with the use of advanced statistics, this is not a completely unreasonable approach: in many ways, this approach attempts to fulfill the reductionist promise of the molecular biology revolution by characterizing individuals by their fundamental generative potential, with the added benefit of potentially facilitating the inference of mechanistic consequences associated with a particular generative potential (i.e., the individual's genome). In certain areas, such an approach is unquestionably beneficial: in the *post hoc* analysis of intragroup variability, itself a vastly underappreciated aspect of experimental and clinical science, this methodology can identify potential members of the treatment group that would be more likely to respond positively to a particular therapy. This procedure is currently most promising in the area of cancer therapeutics, where the molecular profile of tumors is being used to guide the selection of chemotherapy regimens. It is also very important in terms of risk assessment and prognosis, where certain known genetic configurations are recognized as being associated with specifically defined clinical trajectories. In addition to existing implementations to aid in diagnosis and prognosis, genomic characterization is increasingly being used to design clinical trials, with the eventual goal of "personalizing" health care, a concept that is both intuitively appealing and sensible.

However, the problem with this overall concept is that the means by which these genomic data are analyzed is almost entirely statistical in nature, and as a result, correlative. Now, we recognize that correlations are powerful (remember, Ptolemic geocentrism was "predictive" for nearly 2000 years), and they are fine if they are applied to a group fundamentally similar to the group that was used to generate the original data. But these methods do not result in anything more than a *hypothesis* of mechanism, and moving toward manipulation of that mechanism requires taking the next scientific step. So, in terms of "personalizing" a particular treatment regimen, any extrapolation from a general population-level conclusion to the individual case represents a hypothesized effect, one that cannot be evaluated in other way than after its been tried. In short, the end result of this strategy of Personalized Medicine is a series of progressive clinical trials with an $N = 1$, with only the possibility of retrospective analysis.

This, of course, runs counter to the other prevailing theme in clinical medicine we have already noted above, namely Evidence-Based Medicine. At one level, Evidence-Based Medicine sounds very appealing: no practice in medicine will be performed unless it has been scientifically proven, with the standard of proof being the Randomized Prospective Clinical Trial. This gives the impression of objectivity to the practice of medicine, moving it farther away from an ill-defined "art" to something that can provide both patients and the people who pay for those patients' care something more tangible to hang on to. As with all the controversial *cul de sacs* we discuss in this book, there is always a seductive kernel of truth in the proposed approaches. For Evidence-Based Medicine, the fact is that, given the variation in physician practice and quality, evidence-based guideline/protocol medicine is better than the average practitioner (by definition!). The problem, of course, is that no practitioner thinks of himself as "average," and a multiplicity of loopholes exist that allow one to rationalize that a particular piece of "evidence" simply does not apply to him. This rationalization is made easier by all the issues we have seen with Randomized Prospective Clinical Trials above; the set of critiques applicable can easily be used to rationalize why "my patients are different" (again, falling back into biology's historical legacy).

The other clear problem with Evidence-Based Medicine, if one has some historical perspective on the history of science, is that it sounds suspiciously like the program of the Logical Positivists, who if we recall attempted to say that the only valid scientific statements were those that could be proven definitively to be true. As we have seen, their attempt to place stringent criteria on what could be discussed turned out to apply to only a very small, negligible piece of the world. So too is the danger of too rigid an adherence to the tenets of Evidence-Based Medicine, as so infamously pointed out in the notable and oft-quoted article on the efficacy of parachutes in Ref. [10]. In this semi-satirical article, the researchers provide a systemic review of the efficacy of parachutes of reducing mortality in the face of a "gravitational challenge" to an individual. Using the metrics and procedures promoted in the Evidence-Based Medicine community, they demonstrated that parachutes could not be proven to be efficacious in the face of a gravitational challenge, and therefore their use could not be recommended. The authors concluded with the useful suggestion that perhaps a Randomized Prospective Clinical Trial could be performed, participation in which could be determined by the vigor with which the Evidence-Based Medicine paradigm was promoted.

The fundamental issue with clinical trials (and, by extension, Evidence-Based Medicine) is that they are entirely phenomenological: they are essentially the comparison of data in the context of an intervention. This is not to say that there is no inference of mechanism present from the intervention. In fact, there is a long chain of mechanistic inferences underlying the ostensible rationale for the implementation of a particular intervention; this is the chain of reasoning that is ostensibly evaluated at the preclinical level through multiple passages through the Scientific Cycle. But the fact is that the longer that chain of inference, the farther and farther away one gets from a mechanistic cause for any differences present in the data. If this result of this process is a positive, beneficial one, then it does not matter why the intervention worked; one may have an idea or hypothesis for why it might be so, but as a means to guide practice it does not really matter. After all, the statistics do not lie! However, the problem arises when the intervention does not work; in this case, there is a whole set of reasons why this might be so. As we have noted above, there is a ready set of methodological issues that might explain why the data produced the results observed, but we suggest that at a fundamental level, the differences between the scale of resolution of the clinical trial and the presumed mechanism precludes the ability to determine what sort of investigation needs to be done next. The attempt to find resolution to this tangle of possibilities is akin to untying the Gordian Knot; even Alexander's solution of cutting through the knot does not yield any information regarding how it was tied. Since we do not know how the tangle arose, we do not know how to avoid it happening again: rather than Alexander, we are now Sisyphus.

We aim to break that loop of futility. The goal of Translational Systems Biology is to define a process that produces "useful failure." This means that we need to develop a process that combines the principles of Popperian falsification with the Abduction of Pierce in order to learn from failed experiments and trials. This is particularly important with respect to clinical trials, which are completely phenomenological. The current process by which clinical trials are evaluated and the means by which they evolve from the underlying basic science research does not allow for a systemic means of coping with failure. Translational Systems Biology is the application of the concept of Popperain Falsification and Abductive reasoning beyond their current application to individual experiments; rather it is the development of investigatory viewpoint and workflow with a rationale based on sound scientific and philosophical foundations.

The Translational Dilemma is accentuated the problems inherent to both ends of the biomedical research pipeline. It is inevitable that there would be some reassessment of the internal processes at either end (though it is surprising that it was not done as much as would be expected—more on this later in Chapter 2.6. We will next look at one of the most pronounced realizations of this introspection that introduced a potential challenge to the existing reductionist Newtonian paradigm, "complexity."

References

[1] Hausheer FH, Kochat H, Parker AR, Ding D, Yao S, Hamilton SE, et al. New approaches to drug discovery and development: a mechanism-based approach to pharmaceutical research and its application to BNP7787, a novel chemoprotective agent. Cancer Chemother Pharmacol 2003;52(Suppl. 1):S3–S15.

[2] Michelson S, Sehgal A, Friedrich C. *In silico* prediction of clinical efficacy. Curr Opin Biotechnol 2006;17(6):666–70.

[3] Vedani A, Dobler M, Lill MA. The challenge of predicting drug toxicity *in silico*. Basic Clin Pharmacol Toxicol 2006;99(3):195–208.

[4] Vodovotz Y, et al. Translational systems biology of inflammation. PLoS Comput Biol 2008;4:1–6.

[5] Vodovotz Y, An G. Systems biology and inflammation Yan Q, editor. Systems biology in drug discovery and development: methods and protocols. Totowa, NJ: Springer Science and Business Media; 2009. p. 181–201.

[6] Grosser T, Yu Y, Fitzgerald GA. Emotion recollected in tranquility: lessons learned from the COX-2 saga. Annu Rev Med 2010;61:17–33.

[7] Stronks K, Wieringa NF, Hardon A. Confronting diversity in the production of clinical evidence goes beyond merely including underrepresented groups in clinical trials. Trials 2013;14:177.

[8] Raimond V, Josselin JM, Rochaix L. HTA agencies facing model biases: the case of type 2 diabetes. Pharmacoeconomics 2014.

[9] Schiller P, et al. Quality of reporting of clinical non-inferiority and equivalence randomised trials—update and extension. Trials 2012;13:214.

[10] Smith GC, Pell JP. Parachute use to prevent death and major trauma related to gravitational challenge: systematic review of randomised controlled trials. BMJ 2003;327(7429):1459–61.

2.5

Complexity in Biomedical Research: Mysticism Versus Methods

com·plex·i·ty: **noun** \kəm-'plek-sə-tē, käm-\
: *the quality or state of not being simple: the quality or state of being complex*
: *a part of something that is complicated or hard to understand:* **Merriam-Webster Online**

A jumbo jet is complicated, mayonnaise is complex: **Attributed to a Frenchman by Paul Cilliers**

The aim of science is to seek the simplest explanation of complex facts… Seek simplicity and distrust it. **Alfred North Whitehead**

What is complexity? As with many terms that are used in everyday language and have a nontechnical meaning, there is considerable ambiguity and inconsistency when the word is applied in a scientific or pseudoscientific fashion. Bacon categorized this challenge as overcoming the Idols of the Marketplace: confusions due to the use of language and taking some words in science to have a different meaning than their common usage. In the common usage of language, it is perfectly fine for us to adopt a standard of: "I know it when I see it" (à la pornography), but when a word is used in an ostensibly scientific context there is an implied degree of precision that, if not actually met, can lead to troubling interpretations and counterproductive assumptions. We suggest that it is the application of exactly that scientific precision to the issue of terminology that can help us define a scientifically legitimate use of the term and what might not be a valid use.

Precision in meaning requires a clear and concise definition, so that you can tell with as little ambiguity as possible whether or not the term is applicable to a particular subject. In other words, it requires the complete opposite of the "I know it when I see it" standard. With this in mind, we can at least start to compile an admittedly noncomprehensive list of what might be considered legitimate applications for "complexity." For instance, "complexity" has a specific meaning in Theoretical Computer Science, where it is a means of characterizing types and classes of computational problems. "Complexity" also has a specific meaning in the field of Information Theory; Kolmogorov Complexity is a technical term used to describe a property of algorithms used to generate information (which also has a formal definition). Mathematics also has its complex numbers and its associated subdiscipline of Complex Dynamics involving the operation and manipulation of recursive functions; deterministic chaos and fractals are addressed in this area of study. While some of these usages do find application in biomedical research (sometimes appropriately and, as we will see, sometimes not), in general, we will not concern ourselves with their specifics. Rather, the importance is in being aware that they exist, so that if we find ourselves dealing with members from these different disciplines we know that the words we use might mean something completely different to them. Our own work has largely involved scientific questions that require the integration of different areas of expertise [1,2], and the issue of clarifying terminology is a central one in this type of interdisciplinary setting.

In the last quarter of the twentieth century, scientists across several disciplines recognized that their traditional (and previously successful) approaches to scientific inquiry were somehow insufficient or inaccurate when applied to an increasingly significant aspect of their areas of study. When these experts from different fields interacted, they identified similarities in some of the types of problems they were experiencing, problems that they had previously thought were specific to their individual fields (see our earlier discussion of the natural tendency of scientists to categorize and hence make distinctions). Given their scientific bent, they sought to formulate principles about this class of problems. Since all (or at least the vast majority) of these disparate disciplines developed out of the Western Scientific Tradition (summarized in Chapter 2.1), this led to a reassessment of some of the fundamental components of the scientific method. While this unquestionably occurred concurrently among different groups in different places, perhaps the most famous of these convergences occurred in the 1980s with the formation of the Santa Fe Institute.

The Santa Fe Institute was founded in 1984 by a group composed primarily of physicists, with the intent of promoting interdisciplinary research aimed at addressing this set of similar problems identified at the cutting edge of several different fields of study, including biology, economics, ecology, and neuroscience. Most of these problems arose because standard investigatory procedures were unable to reliably characterize known phenomena generated by the interactions among large numbers of components, thereby producing a knowledge gap between what was thought to happen at the micro level and what was observed at the macro level. The concept of "Complexity Science" and various adjectival applications of "complex" were used as umbrella terms to characterize these sets of problems. At the time, the concept of research scientists operating outside their area of specialty was relatively novel. As we will see in our discussion about the academic environment and its impact on scientific progress, the structure of academia fostered (and to a great degree continues to foster) siloed and compartmentalized scientific communities with limited interaction among specific disciplines. This, in fact, was also something Bacon warned about back in the seventeenth century with his description of Idols of the Theater and the restrictions of Academic dogma. Interestingly, being by and large physicists, the founders of the Santa Fe Institute did not look to their own discipline as an area that needed enlightenment!

The fact that there appears to have been an impetus for a group of theoretical physicists to believe that they could impart insight into other areas of study is both characteristic and telling, with both positive and negative consequences. The positive effects of the initial manifestation of the Santa Fe Institute are unquestioned: (i) it introduced the tools and methods of advanced applied mathematics to areas where they had not been considered, much less used; (ii) fostered the concept of multidisciplinary research; (iii) offered the possibility of theoretical and abstract inquiry into previously empirical sciences (most notably biology, ecology, and evolution); and (iv) introduced the notion of mathematical modeling and simulation to these fields, i.e., *in silico* experimental investigation. Unfortunately, there were also negative effects (though all almost certainly quite inadvertent): (i) it perpetuated the perception that physics was the "parent" science, through which all other disciplines could be reduced; (ii) it imparted the reliance seen in physics on known (or at least highly trusted) fundamental laws to other disciplines; (iii) it introduced the physics notion of the goals of modeling and simulation, specifically in terms of the standard of prediction, to other fields of study; and (iv) it fell into the trap of not having a precise enough definition of "complexity." We address some of the negative aspects #1–3 in other sections of this book (Chapters 3.2, 3.3, 4.3, and 4.4); here we will discuss the negative fallout from the terminology problem.

Perhaps part of the problem was that while early researchers in "Complexity Science" were, by and large, highly accomplished scientists, they also had a bit of the poet in them. Such terms and phrases as: "emergence"(mysterious), "edge of chaos"(exciting), "self-ordered criticality"(intriguing), "scale-free"(who doesn't want something free?), "power laws"(powerful), and "complexity" itself (well, complex…), roll easily off the tongue and are able to capture the imagination. This aesthetic appeal of these terms and phrases, combined with an imparted sense of universality inherent to their multidisciplinary application, lent these terms and their associated investigations an air of epiphany and revelation. As we have noted earlier, the idea that we can experience an "Aha, now I see the light" moment that brings order and explanation to our world is something buried very deep in our brains, and if, given the imprimatur of science, it is very seductive. As a result, some legitimate technical terms (such as "self-ordered criticality," an identified property of dynamical systems) became misapplied, others (such as "scale-free" and "power laws") took on a transcendental quality, while others, such as "emergence" and "complexity," just could not be defined, thereby falling into the "I know it when I see it" class of "definitions."

This last group of terms is the most troubling and disturbing; the misapplication of a technical term can be corrected by a reorientation to its specified usage (see our prior list of scientifically legitimate uses of "complexity"), but when you are not quite sure what something means, then that presents a significant problem in a scientific endeavor. Recalling that the fundamental property of science is doubt, and that, as a result, the Popperian concept of falsification is an operational goal of science, the applications of a term to specific topics must be able to be challenged: i.e., an individual must be able to challenge its appropriateness to a specific case. Therein lies the primary difficulty with poorly defined terms such as "emergence" and "complexity": how can one decide when the term is not being used appropriately? Without the ability to exclude an application or falsify a conclusion, the field moves from science into belief. This, in fact, is what happened to "Complexity." Not that we should always consider Wikipedia as a primary authority, but it is telling that the Wikipedia page for "Complexity" as of the Spring of 2014 comes with the header: "This article **may need to be rewritten entirely to comply with Wikipedia's quality standards.**"

(Bold emphasis from the original.) An investigation of the Wikipedia Talk Page provides an illuminating view of the issues facing the use of "complexity" in a potentially scientific context.

The issue is that the terms "complexity" and "emergence" (which became a primary property used to define "complexity") took on an almost mystical tone. Invoking "complexity" and "emergence" became a way to invoke a higher order, implicitly beyond human understanding, in the description of systems and their behavior. The pedigree of these terms, arising from a group of theoretical physicists (the closest thing science has to a high priest class), lent intrinsic scientific credibility to such pronouncements. Not only did this lead to the creation of a scientific patina to subjects that should not have been so colored, the use of the term "complexity" could divert intellectual inquiry away from established and rigorous disciplines. For instance, "complex systems" are often characterized as having nonlinear dynamics, having multiple feedback loops, and being "open" systems. The next question, then, must immediately be: what separates the "discipline" of "Complexity Science" from the actual rigorous study of Nonlinear Dynamical Systems, or the application of Robust Control Theory, or the field of Dissipative Nonnegative Equilibrium Systems [3]? It has unfortunately reached a point where the mere inclusion of "complex" or "emergence" in a topic heading will lead to derision and immediate dismissal by exactly the quantitative disciplines that should be aiding in the investigation of these phenomena.

But, despite all this, "complexity" is such a potentially useful word that it just cannot be discarded. It does seem to fit particularly well with how we would like to characterize some of the critical issues that we, in the area of biomedical research, would like to address. Therefore, rather than abandon the term completely, we will, for our purposes, attempt to construct a way that we can use "complex" in our discourse without all the negative baggage this term has accumulated over the past few decades. To start with, we can define what "Complexity" should not be:

1. *It should not be a replacement for ignorance.* It cannot be complex just because you currently cannot understand or predict a system's behavior.
2. *It should not represent an intellectual dead end.* You cannot say "it is complex" and then not have a scientific way forward.

As simple as the above rules of engagement appear, they are consistent with the properties of science: there must be a means to ask additional, potentially answerable questions. So, what properties should a system have in order to allow the use of the term "Complex"? We propose the following:

1. *It is a dynamic system.* This means that its characterization requires accounting for the dimension of time. Therefore, this characterization also requires a mechanism by which to generate the causal chains defining behavior as the system moves through time.
2. *It involves feedback loops, both positive and negative.* A system, by definition, involves multiple components that interact in some fashion. For us, in a complex system these interactions take the form of feedback loops, either negative feedback (which provides control) or positive feedback (which provides responsiveness). Feedback relationships necessarily lead to nonlinear behavior, a property often invoked in the description of complex systems.
3. *It is a multiscale system.* This means that there is some degree of recursion in the system's components, such that the properties of dynamic behavior and feedback loops exist in a nested fashion. Practically, this means that the system can be characterized at different levels of organization, where it is possible to "hide" the inner workings operating at a lower level of organization.

We completely understand and accept that Criteria 1 and 2 describe nonlinear dynamical systems and the components of Robust Control Theory. In fact, we wish to use these criteria to link the tools and methods from those disciplines explicitly to our set of problems in the biomedical sphere. Rather than create a parallel track of our own "Science," our goal is to leverage the recognition that such expertise exists to enhance the overall progress of the primary goal of biomedical research: enhancing human health (hence, Translational Systems Biology). An additional appeal of being able to use the word "complexity" occasionally goes back to the origins of the Complexity Science movement in the 1980s: it points to a clear need to change the existing standard research paradigm, and emphasizes that there are significant limitations to the traditional research process. The origins of Translational Systems Biology have the same impetus, being derived from recognition of the limits of, and associated frustrations with, the effectiveness of traditional biomedical research to create new therapeutic modalities. However, we hope to distinguish ourselves from some of the ill-fated manifestations of "Complexity Science" by proposing a specific research agenda, involving the integration of methods drawn from a range of advanced quantitative and theoretical fields of study into a truly multidisciplinary approach.

II. THE CURRENT LANDSCAPE: WHERE IT CAME FROM, HOW WE GOT HERE, AND WHAT IS WRONG

Since we are working in the biomedical arena, we also need to address the term "biocomplexity." In addition to all the caveats concerning the misuse of "complexity," biocomplexity is, to some degree, a redundant term if it is used to describe a biological system. It should be evident that, given our description of "complex" above, that all biological systems fall into this category. However, nothing is ever simple (or should we say things are always complex!). One accepted use of biocomplexity in biomedical research is in the discussion the characteristics of biological time series data, most often with respect to physiological signals such as heart rate, blood pressure, or respiratory rate [4–6]. In this context, the term "biocomplexity" often refers to the Information Theory usage, being incorporated into metrics used to characterize the amount of multidimensional regularity seen in various physiological time series data. When we discuss physiological signal processing and interpretation as a means of gaining insight into the physiological state of patients, we will specifically note how we are using the term. For the most part, when we apply "complex" as an adjective throughout this book, it will be in order to demonstrate that we have reached the limit of traditional reductionist and linear experimental methods. Implicit in our use will be the recognition that we will need to apply some of the approaches initially proposed by the researchers at the Santa Fe Institute: (i) application of methods from the fields of nonlinear dynamical systems, robust control theory, and computer science; (ii) abstract and formal representation of the system; and (iii) the use of modeling and simulation to leverage existing mechanistic knowledge to create new knowledge. We add to these strategies the explicit goal of pursuing biomedical knowledge to facilitate its application in the clinical setting, with a specific emphasis on providing a rational pathway to develop therapeutics [7]. However, we recognize that accomplishing this task not only needs to overcome the scientific and methodological challenges inherent with dealing with "complex" problems, but also it will be necessary to overcome human factors manifesting in psychological, social, and organizational barriers (Bacon's Idols of the Mind) that helped produce the silos that currently define the scientific community and hinder the necessary shift toward a new research paradigm. This is not only true as individual researchers try and overcome existing prejudices and groupthink, but also in terms of developing the types of multidisciplinary research teams that grew out of the initial studies of Complexity. We will see just how pervasive and embedded these barriers are in the following section. As is nearly always the case, having foreknowledge is a means of being forewarned as we develop a strategy to overcome these barriers.

References

[1] Vodovotz Y, Clermont G, Hunt CA, Lefering R, Bartels J, Seydel R, et al. Evidence-based modeling of critical illness: an initial consensus from the Society for Complexity in Acute Illness. J Crit Care 2007;22:77–84.
[2] An G, Hunt CA, Clermont G, Neugebauer E, Vodovotz Y. Challenges and rewards on the road to translational systems biology in acute illness: four case reports from interdisciplinary teams. J Crit Care 2007;22:169–75.
[3] Csete ME, Doyle JC. Reverse engineering of biological complexity. Science 2002;295(5560):1664–9.
[4] Godin PJ, Buchman TG. Uncoupling of biological oscillators: a complementary hypothesis concerning the pathogenesis of multiple organ dysfunction syndrome. Crit Care Med 1996;24(7):1107–16.
[5] Goldberger AL, Rigney DR, West BJ. Chaos and fractals in human physiology. Sci Am 1990;262(2):42–9.
[6] Dick TE, Molkov Y, Nieman G, Hsieh Y, Jacono FJ, Doyle J, et al. Linking inflammation and cardiorespiratory variability in sepsis via computational modeling. Front Physiol 2012;3:222.
[7] An G, Bartels J, Vodovotz Y. In silico augmentation of the drug development pipeline: examples from the study of acute inflammation. Drug Dev Res 2011;72:1–14.

Suggested Additional Readings

Note to readers: As the reader can hopefully discern from the content of this chapter, the exiting "literature" on Complexity, particularly in terms of it treatment in popular science books, needs to be read with a scientifically critical eye. This topic is well suited to colorful and eloquent exposition: the readers should consider themselves warned. The texts below represent a survey of this topic mostly from the early heyday of Complexity in the 1990s, which reflect the optimism in which this topic was introduced.

The Third Culture: Beyond the Scientific Revolution by John Brockman

Complexity: The Emerging Science on the Edge of Order and Chaos by M. Mitchell Waldrop

Complexity: A Guided Tour by Melanie Mitchell

Fire in the Mind: Science, Faith and the Search for Order by George Johnson

Foundations of Complex-System Theories: In Economics, Evolutionary Biology and Statistical Physics by Sunny Auyang

Order out of Chaos by Ilya Prigonine

Catastrophe Theory by Vladimir I. Arnol'd, G.S. Wassermann (Translator) and R.K. Thomas (Translator)

Complexification: Explaining the Paradoxical World Through the Science of Surprise by John Casti

2.6

Human Nature, Politics, and Translational Inertia

It was six men of Indostan
To learning much inclined,
Who went to see the Elephant
(Though all of them were blind),
That each by observation
Might satisfy his mind.
The First approached the Elephant,
And happening to fall
Against his broad and sturdy side,
At once began to bawl:
"God bless me! but the Elephant
Is very like a WALL!"
The Second, feeling of the tusk,
Cried, "Ho, what have we here,
So very round and smooth and sharp?
To me 'tis mighty clear
This wonder of an Elephant
Is very like a SPEAR!"
The Third approached the animal,
And happening to take
The squirming trunk within his hands,
Thus boldly up and spake:
"I see," quoth he, "the Elephant
Is very like a SNAKE!"
The Fourth reached out an eager hand,
And felt about the knee
"What most this wondrous beast is like
Is mighty plain," quoth he:
"'Tis clear enough the Elephant
Is very like a TREE!"
The Fifth, who chanced to touch the ear,
Said: "E'en the blindest man
Can tell what this resembles most;
Deny the fact who can,
This marvel of an Elephant
Is very like a FAN!"
The Sixth no sooner had begun

About the beast to grope,
Than seizing on the swinging tail
That fell within his scope,
"I see," quoth he, "the Elephant
Is very like a ROPE!"
And so these men of Indostan
Disputed loud and long,
Each in his own opinion
Exceeding stiff and strong,
Though each was partly in the right,
And all were in the wrong! **"The Blind Men and the Elephant" by John Godfrey Saxe in a retelling of an ancient** **Indian parable**

Everyone is entitled to his own opinion, but not his own facts. **Daniel Patrick Moynihan**

That smooth-faced gentleman, tickling Commodity.
Commodity, the bias of the world; **William Shakespeare, as spoken by the Bastard in "King John," Act 2, Scene 1,** **Lines 574–575**

I don't want to be a member of any club that would have me as a member! **Groucho Marx**

We have outlined previously how the scientific endeavor reflects fundamental human traits: the quest for order, curiosity driving the search for understanding, the desire to better one's life. It could be argued that it is the trans-generational acquisition of knowledge, incorporating the transfer of life experiences to successive generations independently from our genetic code that may be the defining characteristic that sets humanity apart from the vast majority of organisms on our planet. But since it is a human endeavor, the practice of science is also subject to human flaws. Those same psychosocial aspects that may have served well in more evolutionarily dynamic times, such as emphasis on the self and one's own tribe, fear of the unknown, distrust of otherness, competition for resources by any means, now manifest in many of the sorrows present in human society: greed, narcissism, inequity, prejudice, ecological and environmental collapse, etc. Accepting that these potential traits are embedded in our evolutionary heritage does not excuse their antisocial manifestation today, but rather helps frame the scope of the challenge we face if we are to overcome them as part of the social contract under which we all live.

So, is it recursively hypocritical of us to repeatedly refer back to Francis Bacon's Idols of the Mind? After all, by using his list of Idols as a guide toward good scientific practice we violate some of tenets warned against by the Idols' existence. Of course we can rationalize that this list *really* does reflect human nature, but then we would be making the same mistake that centuries of geocentric astronomers made building upon fundamental principles they also *knew* to be true. The best we can do to break the cycle of rationalization is to adopt a scientific approach focusing on the collection of evidence from our observations of the world and seeking to disprove Bacon's list if we can.

Unfortunately, the evidence provided by history thus far cannot falsify Bacon's characterization of the barriers to scientific thought. The history of science is littered with tales of innovators who passed from this world before their foresight could be completely understood. At best they were underrecognized and underappreciated, and at worst they were ostracized and hounded. It is ironic that so many of those individuals, often most admired historically as reflecting the "best" of human curiosity, ingenuity, and creativity, so frequently suffer within the context of their own contemporary communities. This trope is so common that it reaches the level of a truism: the unrecognized genius. It is therefore somehow odd, given its amazing frequency, that each successive society, clearly aware of the lessons of the past, continually renews the same error. Moreover, one would think that by now someone would have figured out a way to engineer a systematic strategy around this barrier. Let us make an attempt...

SETTING THE TABLE WITH BACON

We have already placed our eggs in Bacon's basket, so we start our investigation of the roots of the barriers to advancement of scientific thought by examining the basis of his list of Idols and describing how they contribute to the current barriers facing the biomedical research community.

Idols of the Tribe (Idola tribus): This refers to the tendency of humans to perceive more order and regularity in systems than truly exists, and is due to people following their preconceived notions about how things are and how things work.

This Idol points to an inherent conservatism in human kind, a fear of the unknown that leads to a holding on of existing concepts and worldviews. Evolutionarily, this makes perfect sense: if something has served us well in the past, or even not so well but has been adequate, then the known condition ("the devil we know") provides greater comfort and security than the unknown ("the evil we do not know"). Innovation is inherently disruptive, and historically and evolutionarily disruption and disorder have represented danger. This survival instinct is buried in our genes, and therefore places a considerable barrier to innovation. There is, however, a persistent potential tension that exists between the urge to maintain the *status quo*, and the desire to innovate. This tension arises from a cost–benefit analysis every person executes, at least at an unconscious level. Two general circumstances can affect the balance between the *status quo* in favor of the unknown over the known: (i) the known condition is so bad, so inadequate, that it cannot be considered to be a sustainable option, and (ii) the negative consequences of innovation are so blunted, as to minimize the risk of innovation and allow the freedom to speculate. Interestingly, these two conditions appear at opposite poles of societal stability. For the vast majority of human history, the First Condition has been the impetus for innovation and discovery: as a matter of survival, people, needed to pick up and move along, find new worlds and new solutions. This strategy can be termed "Necessity is the mother of invention." In the context of this book, this means that we are required to point out the futility of the current investigatory paradigm, both in terms of methods and structure. The Second Condition is a bit trickier: this circumstance presupposes that the particular society is capable of producing at least a basal level of security and prosperity such that intellectual efforts do not need to be entirely focused on survival. This allows at least some members of the society to explore new ideas that, while they may not have an immediate payoff, can advance the general state of knowledge and set the stage for future advancement. A cursory historical assessment can identify several such periods in human history: Athens in the Age of Pericles, the Golden and Silver Age of the Roman Empire, the flowering of Islamic learning from the seventh to the ninth centuries, the Han and Ming Dynasties in China, and the Age of the Victorian Gentleman Scientists in the nineteenth century. However, the same societal stability that allows the luxury of such intellectual pursuits also provides barriers to the adoption of the resulting innovations, and all too often the advances seen in these settings are only recognized in retrospect. It could be argued that our current society rests much more in the Second Condition than the First, though we hope to make the case that in actuality we are closer to the edge of an unacceptable *status quo* than we would like to think.

> Idols of the Cave (*Idola specus*): This is due to individuals' personal weaknesses in reasoning due to particular personalities, likes, and dislikes.

Some may consider humans to be rational creatures, but unfortunately history, society, and science suggest rather that humans are primarily *rationalizing* creatures. Human beings overwhelmingly emphasize accumulated evidence that supports their existing mindset; we are in actuality far from the coldly empirical, skeptical creatures idealized by Hume and the Logical Positivists. To a great degree, the Scientific Method is intended to counteract this tendency, an attempt to place objective criteria and constraints on how we interpret our observations of the world and the conclusions that we make. But, as Bacon realized, we are ultimately humans with human failings, and therefore subject to falling before this particular Idol of the Cave. Challenging this Idol presents a particularly difficult task, since perhaps more than any other this one comes closest to an individual's sense of self. It should be noted that the strategies of worshiping the Idol of the Cave are often invoked in the pursuit of the other Idols in Bacon's list. Again, the hope is that discourse can be framed in such a way that there is some baseline level of agreement from which to build; from that point discussion and disagreement can be resolved through a mutually agreeable process. However, as we can unfortunately see in multiple aspects of human interaction, particularly today, we are unable to meet even the basic criteria suggested by former US Senator Daniel Patrick Moynihan, who said: "Everyone is entitled to his own opinions, but not his own facts." Today, all too often discourse has devolved into a dispute about facts, often without a realization about how facts are actually obtained. Perhaps this condition can be considered a symptom of humanity's growing pains: as we move from the childhood of certainty offered by monolithic authority (Church, State, etc.) into adolescent, postmodern awareness of the shakiness of those foundations. The advent of the Internet and, more recently, the explosion of social networking, with its promise of instant connectivity but with the pitfall of promoting narcissism, have only made these problems worse. Discourse often now feels like a shouting match, one in which rationality is in short supply and the least common sense is common sense. And a key pillar of scientific discourse, that of peer review, often fares little better.

Postmodernism, because it disrupts prior concepts of authority and leads to a relativistic, subjective view of the world, is often misinterpreted as "anything goes." This is akin to the mindset of an adolescent, where developmental awareness and hormone surges lead to an overturning of parental authority, often accompanied by existential angst that since everything in the world can be challenged, none of it can have real meaning. In order to mature and progress past this phase we, as adults, need to develop grounding strategies to orient us and allow us to operate in the world. Science, ostensibly tasked with identifying facts, is no different. On one hand, this is a good thing: as Hume and Popper tell us, we should always distrust our "facts." However, how those facts are defined and arrived at should also be subjected to the same rigorous discourse as the specific facts themselves. This is why we ground our discussion of the next steps for biomedical research with historical and philosophical background: it will allow us to separate those manifestations of pseudoscience that take advantage of the instabilities generated by the relativism brought about by postmodernism (Creationism/Intelligent Design, mystical manifestations of Complexity, etc.) from those paths that might be considered legitimate successors to the scientific tradition and honor its basic, foundational tenets.

> Idols of the Marketplace (*Idola fori*): This is due to confusions in the use of language and taking some words in science to have a different meaning than their common usage.

One of the origins of postmodernist relativism can be found in the deconstruction of language. We had introduced Ludwig Wittgenstein earlier as the author of *Tractatus Logico-Philosophicus*, the work that was so influential to the Logical Positivists; here, we return to Wittgenstein for the second phase of his career, in which he essentially—disruptively?—overturned many the primary tenets of his *Tractatus*. After an interval as a schoolteacher in rural Austria, Wittgenstein returned to academia at Cambridge; the product of the later phase of his career would be published posthumously as *Philosophical Investigations*. In this work, Wittgenstein focused on the meaning and use of language, with the resulting conclusion that language was not intrinsically tied to any objective reality, but rather represented a game being played among the users of language. Meaning now became fluid and dependent upon the output of the interactions among the engaged parties. We have touched upon this phenomenon in the earlier discussion of "Complexity" in Chapter 2.5 and how this term has been used and abused. We have also noted that different domains may use the same word to mean different things, all valid within their particular context, but leading to considerable misunderstanding in cross-disciplinary encounters. One potential adverse outcome from an initial encounter would be if neither party is willing to budge on how potentially ambiguous terminology could be used, with each party obstinately maintaining the position that the other party is simply wrong. Such a situation would obviously form a considerable barrier to future discourse. A primary lesson of Wittgenstein's concept of language as a game is that the rules of the game must be established as early as possible; hence our emphasis on trying to establishing definitions for specific contexts at the outset. Given the fluidity of language, we suggest that the recurrent reorientation of the meaning of a particular term and how it is being used is an absolute necessity for multidisciplinary encounters, and as a particular field progresses.

We relate our own experience with this phenomenon in a report published in *Journal of Critical Care* concerning the initial formation of the Society for Complexity in Acute Illness [1]. The formation of this Society involved an attempt to form collaborative relationships among experimental biologists, clinical researcher, applied mathematicians, and computer scientists. Early conversations invariably involved negotiations around the use of what had been thought to be unambiguous terms; for instance, the concept of "good data" occupied nearly a day's worth of discussion (the mathematicians, given their sense of precision, could not understand why the biologists were showing them data with error bars, providing the comment that if there was an "error" then surely something was not done correctly!) [1].

In addition to dealing with ambiguity, the alternative case is all too common in today's scientific environment, in which trendy jargon is used to give the impression of timeliness. Examples of this include the use of such currently popular catchphrases as "complex," "novel," "systems-oriented," "integration," "network," "Big Data," and "synthesis," just to name a few. These terms are sprinkled into manuscripts and proposals, providing the patina of these concepts prior to the presentation of business as usual. Here the "game" is misdirection and obfuscation; by attempting to leverage currently popular memes, individuals are able to repackage and resell the same old product.

Of course, we open ourselves up to the same critique, because the fact of the matter is that we find ourselves with the dilemma we described in our discussion of "complexity" in Chapter 2.5. Should we allow the misappropriation of these terms to prevent our ability to communicate our own interpretation? We would argue that by recognizing the limitations of our terminology we have gone a considerable step toward overthrowing the barriers formed by the various Idols: by recognizing that there is no philosophical or scientific basis for *de facto* authority, we force the requirement of being explicit and specific in the definition of terms, and therefore establish a firm grounding for additional scientific discourse.

Idols of the Theatre (*Idola theatri*): This is the following of academic dogma and not asking questions about the world.

Humans are social creatures. We have survived to evolve in competition against stronger and faster creatures in part because of our intellect, and also because we have utilized that intellect to leverage the power of our groups. The roots of group membership run very deep in our being, and are yet another fundamental aspect of our psyches. As we have noted above, what may have been a positive survival mechanism in the past now has its negative consequences. The distrust of otherness and the exclusionary forces of tribalism continue to manifest at every level of society. Academia is no different, particularly if academia can be thought of as an extension of the prescientific priest class. Having "secret knowledge" has been, throughout human history, a means of creating a separate hierarchy within a society, with those holders of secret knowledge are Druids, Clergy, or Scientists. Membership in that class is highly guarded, and subject to rigorous and protectionist selection; i.e., one must "pay his dues" in order to become a member. Bacon realized that Authority is intrinsically self-perpetuating: its justification for existence relies directly on how it is exercised, i.e., as the reservoir of secret knowledge. We recognize that, to a great degree, the other Idols of the Mind come into play in order to perpetuate the Idols of the Theater. The structures and processes that make up Academia take lives of their own, serving to sustain themselves in a perpetual loop. But, of course, it is more complicated than that. After all, there are definite benefits of Authority/Academia, most important of which is that it provides some guidance, or vetting, or standards, in short, it provides order. Not everyone has the time, ability, or inclination to evaluate everything completely on his own; at some point, there must be some delegation of trust to some group or authority to provide guidance in how to proceed and what should be adopted. Thus, the consequences of breaking the self-perpetuating loop of Academia (and industry, too, as we discuss below) involves assuming risks not only for the potential breakers or innovators but also in terms of potential disruption of the existent beneficial roles that Academia can provide.

We see in the numerous origins and effects of Bacon's Idols of the Mind the scope of the challenge we face in attempting to apply the Scientific Method to the scientific community itself. Bacon initially intended his Idols of the Mind to represent warnings to the process of interpreting data and formulating hypotheses within his Scientific Method. However, since they are grounded in human nature, the Idols of the Mind permeate nearly all human interactions. We recognize, then, that the Idols of the Mind are not only barriers to an individual's execution of science but are magnified when we analyze the origins and structure of the scientific community itself. The following sections will relate how specific conditions associated with the history of science in the last half century have led to developments, though which may have been initially well intentioned and reasonable, have unfortunately since fallen under the sway of the Idols of the Mind.

AN EMBARRASSMENT OF RICHES

It is difficult to view the state of the science at the start of the twenty-first century with anything other than an overwhelming sense of accomplishment. In addition to the pure advancement of knowledge across multiple disciplines and progress in technology that a half century ago could only be imagined in science fiction, the scientific community has also evolved to an unprecedented level of scope and sophistication. The rapid acceleration of the dissemination of knowledge, and the attendant effects on education in the sciences, is an incredible success story. It would not be a reach to suppose that there are more trained scientists per capita today than at any other period of history. More is known about more things than ever, and the rate of accumulation of facts is also likely to continue growing both in breadth and in depth. Furthermore, we have generated completely new areas of research such as synthetic biology and brain–machine interface research, for which we have generated brand new facts and all new rules. However, it can easily be seen that this embarrassment of riches carries with it an inevitable consequence: if we know both more about things and about more things, then it is impossible to know everything about everything. We need to strike a balance between knowing a lot about a few things, or a little about many things. We know how this balancing act has turned out (spoiler: we think it is something that needs to be corrected). However, in keeping with our concept that most trends start with reasonable people doing reasonable things, let us look for a moment as a reasonable rationale for choosing one path versus the other. As we have previously noted, the process of science arose to meet the human desire to find order in the natural world. That order implies some underlining process or force that binds together the multiplicity of phenomena we observe, but the investigation of that process necessarily involves a decomposition of phenomena to find out what might be similar. Therefore, the initial step in the investigatory process is one of finding out more about some individual *thing*; it is only after one thing has been characterized that it can be compared to another thing. Even if the initial question is based on the observation that "this thing here is kind of like that thing there," the first step to answering that question requires a breakdown of each of the things into a list of descriptions that can be potentially matched to each other. Given the primacy of this act, it is therefore perfectly reasonable that if faced with the initial choice of delving deeper into a particular subject or looking for new things to examine, the scientific path inherently emphasizes first the deeper inquiry. And so this emphasis on detail-acquiring investigation was mirrored in the developing structures of academia, with the splintering of the general scientific pursuit into different disciplines. The general "Naturalists" of the eighteenth and nineteenth century became chemists, physicists, and biologists as more and more information was acquired and it was increasingly more difficult for a single person to be able to keep up with the cutting-edge developments across all the sciences.

We propose that this phenomenon was accelerated in biology following the molecular biology revolution, due in part to its successes in generating new information, and also due to lingering tradition in biology to emphasize the uniqueness among biological systems. There is an exponential increase in the possible subjects of study once one focuses on biological systems: there are multiple tissues, composed of multiple cell types, containing multiple molecular pathways. Since each pathway represents a very complicated system, its investigation necessarily involves a considerable amount of time and effort. The key here is that given the investigatory paradigm of reductionism, the means of studying a particular pathway necessarily involved *separating that pathway from all its surrounding pathways*! Therefore, subsequent metrics of expertise and accomplishment could be tied to how much detail you acquired about a very specific thing, leading to the truism about the process of basic science, namely that it is a process of knowing more and more about less and less until you eventually know everything about nearly nothing. While this truism certainly exaggerates (some), the fact of the matter is that the structures of academia evolved to fit this paradigm, with increasingly specialized work being done in an environment that was increasingly disconnected from the overall context. Academic accomplishment became tied to the degree of detailed knowledge in a specialized area, because that is how Academic Departments are structured, around specialists themselves; here, we see the self-perpetuating feedback that sustains the Idol of the Theater. *The insidious aspect of all of this is that it makes sense!* Because of course you want to have experts driving investigation, of course you want experts vetting new work, and of course we know that the expansion of science must come after the acquisition of detail. We recognize this need; our point is that we also need to recognize that there is another part of the scientific endeavor that needs to be reemphasized, and that the current structure of the Academic largely prevents this from happening. Let us look at some of the additional societal factors that reinforce this barrier to true innovation.

SHIBBOLETHS IN SCIENCE: THE PROBLEM WITH PEDIGREES

Where did you train? Whose lab were you in? How many papers do you have in *Science* or *Nature*? These are the shibboleths in Academia; how you answer them will determine whether you are part of the "in" crowd, or a poser/wannabe. This situation is a direct outcome of the fracturing of biomedical science we discussed in the prior section: it is a short step to move from the concept of favoring ultraspecialization in study to the valuation of pedigrees of those areas which become known for excellence in that process. Again, this makes perfect sense, and again, this is the insidious nature of the barrier. The rationale for the underlying concept is very reasonable: certain institutions become known for promoting great work, individuals with great capabilities flock to or are recruited to these sites, some of those individuals become recognized a standout scientists, who then themselves become magnets for those with ability. Your academic pedigree therefore becomes a shorthand description of your potential worth and promise; the supposition that exposure to a successful and productive environment will increase your own chances of success and accomplishment. This all sounds very good, and almost certainly can be "proven" by looking at the academic productivity of lineages of researchers from various notable laboratories and investigators. But note that this is actually a rigged game, since one of the primary metrics of evaluation for success is being part of one of these lineages! If the criteria that determines future success is the history of prior success (which was based on a similar process of evaluating potential future success)…*well, it's turtles all the way down.* Is this to say that those fortunate and capable investigators able to claim these valued academic lineages are not doing good and important work? No, this is not the claim. Rather, the point is that, given an environment of limited resources, are those resources being spent on the best ideas and the most capable minds? We posit that we have a reasonable expectation that they may not be due to the inherent aspects of the current structure of the scientific community. And again, it is an intended consequence arising from what would initially seem to be a very reasonable and desirable development: the professionalization of the scientist class.

INCENTIVES AND THE PROFESSIONALIZATION OF SCIENCE

How do the incentives of society influence how science is performed? How do those incentives make us susceptible to the Idols of the Mind? Again, good intentions lead to unintended consequences, and interact together in nonlinear ways that lead eventually to Hell. We have seen the reasonable basis of the fractionation of the biomedical research community, and the resulting importance that fractionation has placed on the pedigrees of its researchers. Now we see how those conditions influence and are influenced by the incentive structures present as a result of the professionalization of science. As with our other examples above, this at first seems like a very good thing: after all who would not want professionals doing a very important task, and being incentivized for doing so? The increased expertise required to deal with and advance the accelerating knowledge across the scientific spectrum cries out for individuals specifically trained and focused on integrating and developing that knowledge; long gone are the days of the Amateur Gentleman Investigators of the Victorian Age. The training of these professionals necessarily requires a similarly specialized environment to train them, mentor them, and send them on their way as fully capable independent investigators. All this makes an incredible amount of sense, does it not?

So what is the problem? The problem comes down to, as it all too often does, money. Because, of course, research takes money: money to pay researchers, money to perform experiments, money to analyze data, money to train trainees. It is not just money to do the research; it is the potential money that comes from being associated with that research, to become one of the Shibboleth-wielding institutions on the inside. Academic institutions, whose sense of self is tied to their ability to host academic productivity (and the extra dollars that come with it), also factor into this equation by reinforcing those structures that most greatly increase their likelihood of getting that research money. As a result, for professional researchers whose livelihoods are dependent upon the continued acquisition of funding, the emphasis inevitably becomes getting and maintaining funding. Certainly there is a ready justification for the pursuit of the research funding: what scientist has not said to him or herself: "I need to get the money so I can continue to do this important work." The question, however, is at what point does the pursuit of the money overwhelm the pursuit of knowledge in the ordering of a researcher's hierarchy of needs? Furthermore, we recognize that this consequence of the professionalization of science folds into it the subject of the prior two sections: the focus on ultraspecialization and the power of pedigrees. If these two properties of the scientific community are the criteria for determining who gets funded (or published in *Science* or *Nature*, which could reasonably be considered en route proxies for funding), then how have we restricted the space for true innovation within the biomedical research sphere? Ultraspecialization is needed because, of course, we want to fund experts; if the goal is to get *some* results (as opposed to potentially impactful results), then this is a fruitful approach. After all, this fits into the existing reductionist paradigm: overall knowledge is being advanced. Of course, this paradigm explicitly devalues the overall impact of a particular project as part of the evaluation process. In a similar fashion, Academic lineages are used as their intended role as shorthand metrics, and form the basis of some sort of prospective assessment of success, which, as we have seen, is a self-perpetuating loop. The third factor here is that those researchers that have been successfully funded are very reluctant to shift gears and focus from what has proven successful in the past. After all, at least in the research-funding world, by design, past performance is a primary determinant of future success; it would be irrational to behave any differently. The unfortunate consequence, however, is that this discourages innovation and exploration, and fosters incrementalism (at best) and obstinacy (at worst); herein we see the manifestation of Bacon's Idols of the Mind to their fullest extent. Challenges to Academic dogma have now moved beyond just scientific discourse. There is now real spending money at stake, as the position of those who would be the experts and arbiters of knowledge is threatened by changes that would diminish their authority. The professionalization of science, particularly in the biomedical arena, has shifted the goal of the job from the acquisition of knowledge to the acquisition of funding. Now, of course, we realize that this is the way it must be, at least for the immediate future; after all, the world does require money and it would be incredibly naïve to imagine that it could be any other way. But as with the other factors we have discussed, recognizing that there are barriers that result from nonremovable components of the scientific endeavor allows us to craft strategies that may potentially mitigate their negative impact. Perhaps we can look outside academia, where the incentives should be more clear-cut. Rather than an Academic structure that hopes indirectly to promote the development of useful science, maybe we can look to the business model, to an industry whose mere existence is dependent upon developing effective therapeutics.

DEEP POCKETS, WITH HOLES: THE PHARMA CONUNDRUM

Or so one would think, as the struggles of Big Pharma in the early twenty-first century are pretty well known, the consolidation within the industry is a mark of decreasing competitiveness because companies are not being as productive as in the past. The decreased efficacy of Big Pharma, to a great degree, is what prompted our development of Translational Systems Biology. The intriguing question, however, is why has the pharmaceutical industry not made the necessary adjustment to fix their fundamental process problems? The answer, of course, is that Bacon's Idols of the Mind are human failings, not limited to the pursuit of science. First of all, let us characterize the primary differences between academic science and the pharmaceutical industry. The most fundamental difference is that biomedical science is (or at least should be) focused on *discovery*, while the pharmaceutical industry is focused on *engineering*. Basic research wants to find out how things work; once we know that then we want drug companies to put that knowledge into usable form through products designed and tested using rigorous processes. Engineering is fundamentally the process of finding a solution given the description of a problem, but in order for this process to be effective, the problem needs to be well defined and, hopefully, circumscribed. Here, we see a major difference between engineering applied to physical systems versus being applied to biology: in physical systems (chemical, mechanical, or physics), the underlying laws of behavior are known and therefore the problem can be defined with constraints based on those laws. Biology is different: we do not have underlying laws (yet), and are operating in an inherently limited space of knowledge, which is ultimately empirical and potentially false. Engineering solutions are able to deal with uncertainties and unknowns, primarily through the use of a succession of abstractions, models, and simulation, but the first step is realizing that this is the case.

Unfortunately, the Pharmaceutical industry has not adopted these time-tested strategies from other engineering domains. To answer why, we must look back to the reliance industry has on Academia, and the incentive structure that is fostered by Bacon's Idols of the Mind. Industry justifiably looks to Academia as its source of expertise; in so doing, industry becomes affected by the same issues and driving forces present within the academic environment: ultraspecialization, emphasis on pedigree, and the resulting groupthink fostered by the Idols of the Tribe, the Idols of the Cave, and the Idols of the Theater. Therefore, the prejudices associated with the evaluation of the science behind the engineering process are carried over into the self-evaluation Pharma performs. Now, add to this the dynamic of diminished expectations: as we have noted, the failures of Big Pharma are well known, including the abysmal translation rate of drug candidate to approved therapeutics. One would think that this should set the stage for overcoming the Idols of the Tribe, as we mentioned earlier: this is now a situation in which things have become so bad that the risk of moving ahead outweigh the consequences of standing pat. Unfortunately, this is not the case, and the reason is that Big Pharma represents large bureaucratic organizations subject to all the internal barriers generated by Bacon's Idols: the operational context for weighing risk of innovation versus benefit is not the wider community, but rather the embedded organizational structures and barriers within the organization itself. And when there is no real expectation that something is going to work, then the criteria for failure becomes moot. In this case, innovation is seen as both risky and expensive, and since the negative consequences of following standard operating procedure are obviated by expectation, then maintaining the *status quo* becomes the path of least resistance. Of course, at some point the music will stop, and a chair must be found, but rather than dealing with fundamental process change, the pharmaceutical industry has adopted business-solution approaches in the form of consolidation and the off-loading of research and development to smaller biotech startups. We argue that the mere fact that these strategies have become necessary for Big Pharma points to the crisis present at the fundamental process level.

Moreover, this situation sets in motion a self-defeating cycle and downward spiral. In Big Pharma nothing will happen because all of the forces at play are stacked against anything disturbing the *status quo*: the company's management is typically conservative, the group leaders are politically savvy and worried about their jobs, the federal government does not want to interfere, and taxpayers are completely oblivious to all of the above. Moreover, some of them may actually believe that, left to their own capitalist devices and free of the fetters of regulation, a flood of new drugs will hit the market. As we state above, though, those very same market forces and the self-sustaining holding pattern that characterize and drive the pharmaceutical industry in fact assure that little innovation will occur. Value will be added to the stocks of pharmaceutical companies, not through new products but through copycat drugs, mergers/acquisitions, and layoffs. This cycle will be repeated, and is unlikely to be interrupted even if companies recruit top management from academia: after all, the forces in academia are essentially the same.

In the midst of chaos, there is also opportunity. **Sun Tzu,** in *The Art of War*

Thus far, we have painted a pretty pessimistic view of the state of the biomedical research endeavor. To harken back to the opening lines of *A Tale of Two Cities*, we do appear to be, paradoxically, at a point where we see the best that biomedical science can offer but are at an appalling level in terms of being able to use that knowledge. We propose that there is a way to build a bridge between these two cities, one that is similarly grounded in the same fundamental aspects of science, history, and philosophy that have led to the current situation, and therefore can be thought of as the next natural evolutionary step in the advancement of biomedical research. We will describe this approach, Translational Systems Biology, in the subsequent chapters, and present examples of how we have developed and applied its tenets to increase understanding of the process of inflammation in both health and disease.

Reference

[1] Vodovotz Y, Clermont G, Hunt CA, Lefering R, Bartels J, Seydel R, et al. Evidence-based modeling of critical illness: an initial consensus from the Society for Complexity in Acute Illness. J Crit Care 2007;22:77–84.

TRANSLATIONAL SYSTEMS BIOLOGY: HOW WE PROPOSE TO FIX THE PROBLEMS OF THE CURRENT BIOMEDICAL RESEARCH LANDSCAPE

3.1

Towards Translational Systems Biology of Inflammation

'...*My name is Ozymandias, king of kings:*
Look on my works, ye Mighty, and despair!'
Nothing beside remains. Round the decay
Of that colossal wreck, boundless and bare
The lone and level sands stretch far away. **Ozymandias, by Percy Bysshe Shelley**

Translation is the art of failure. **Umberto Eco**

So where do we, as the biomedical research community, stand at this point, in the early part of the twenty-first century? It may be tempting to look at what has been accomplished, and what is currently being accomplished, with a sense of satisfaction and confidence. After all, as we have seen, there is much to be satisfied about: since the advent of molecular biology, the essential "Code of Life" had been broken, culminating in the sequencing of the Human Genome. Data can be collected at a level unprecedented and inconceivable even a few decades earlier. An ever-accelerating increase in computing capability has already outstripped Moore's prediction, and seems not likely to slow down for the foreseeable future. But, as we hope we have also demonstrated, there are significant storm clouds on the horizon. The steady reduction in the number of effective therapeutics coming to market is very real, and the relationship between that decrementing trend and the increasing expenditures on basic biomedical research only accentuates the scope of the problem. Pharmaceutical companies are scaling back, if not outright abandoning, their own research programs, and rolling the dice on startups that may have some promising drug candidates, which in turn often get their technology leads from academia. This is a financial strategy born of futility, essentially attempting to outsource their intellectual burden and risk; this is especially worrisome in the current abysmal climate for biomedical research funding. And the more we learn about the diseases that trouble us today, the more complicated and more intractable to the "standard operating procedure" they appear. We do not want to wind up like Shelley's *Ozymandias*, our primary legacy the wreckage of possibility and pride.

How to make sense of this situation? As part of the medical community, we have practiced that time-honored art of diagnosis on the Translational Dilemma, and have come up with a prescription that can fulfill the proscription that "doctor heal thyself," only now applied to the biomedical research community at large. The process of diagnosis was laid out in Section 2, and here is a list that summarizes our findings:

1. The primary diseases that trouble us today, many of which involve *inflammation*, are resistant to the typical, *reductionist* application of the Scientific Cycle.
2. The primary manifestation of that resistance is the *unexpected failure*, at the Clinical Trial level, of candidates that should, by all accepted metrics, have been efficacious.
3. *Big Data* approaches have provided incredible insights into the underlying biology of many diseases. However, the position of Big Data in the Scientific Cycle, i.e., as identifying correlation, does not allow these algorithms to test the mechanistic hypotheses they aid in discovering.
4. A side effect of Big Data is that it has drastically and exponentially increased the set of possible mechanistic hypotheses that could exist, with the consequence that the traditional means of testing hypotheses, experiment, *simply cannot keep up*. This has produced a profound imbalance in the Scientific Cycle.

Translational Systems Biology.

5. Fundamentally, science, and biology in particular, has emphasized finding out ever more *detail* about processes under study. However, the application of this process using current tools, specifically high-dimensional gene and molecular characterization using "omics" technologies, leads to the condition seen in #4. This emphasis on collecting detail is part of biology's historical legacy, and diverts effort from finding out how biological systems function.

6. *Physics Envy* has both benefitted and harmed molecular biology. On the positive side, the desire to emulate physics has led to the idea of applying mathematical tools to characterizing biology, a process of invaluable import. On the negative side, however, this envy has perpetuated the reductionist paradigm, which while suitable for physics and chemistry, does not appear to apply to biology. The persistence of this concept has led to unrealistic targets and investigatory rabbit holes for the application of mathematical and computational methods to biology.

7. The *professionalization* of science has had the unintended consequence of misdirecting intellectual energy into traditional, comfortable, but not particularly effective, practices and procedures, such as those historical artifacts mentioned in #5.

8. Another negative consequence of the professionalization of academia has been the *reinforcement of boundaries* among researchers, labs, and fields of study, limiting the ability to leverage intellectual capital and expertise.

9. The *fragmented* structure of the biomedical research community mirrors the intellectual and investigatory barriers in moving through the drug development pipeline.

10. There are significant practical and methodological constraints on the current scientific practice and procedures in all aspects of that pipeline.

It is said that knowing the problem is half the battle. We disagree. We do not think that the division of effort is well reflected in that statement, but that the scope of the challenge is more accurately reflected in a quote from Winston Churchill: "*Now this is not the end. It is not even the beginning of the end. But it is, perhaps, the end of the beginning…*" We believe that recognizing the problem is "the end of the beginning" of dealing with the problem. Now comes the hard part: designing and implementing a solution.

It is tempting to fall back on technology. Technology has always bailed us out, and it is not unreasonable to believe that human ingenuity, which has not really let us down yet, will continue to rise to meet any bumps in the road we might experience. This, in fact, is a fairly pervasive thought process today, a belief that the brute force application of technological advancement of the tools and methods being utilized today will be the solution. Some examples of this type of thinking include: just get more data, just have faster computers, just develop more detailed simulations, just find the right biomarker/therapeutic target. The popularity of shifting paradigms to Big Data Science with concurrent obituaries for hypothesis-based research reflects a hopefulness that, if given enough data and enough computing power, the answers will emerge.

We respectively disagree. We believe in the Scientific Cycle, and that *all* of its phases are critically and equally important. Our diagnosis of the Translational Dilemma has pointed out that the current Scientific Cycle is imbalanced. Our solution rests on the rational, targeted utilization of technology to restore the balance of a healthy and productive Scientific Cycle. More specifically, our solution explicitly targets the process of getting better therapeutics to the bedside, and is aimed at providing solutions not currently met by existing research strategies, approaches, and domains. This is why we developed Translational Systems Biology, and why we have written this book. We have previously introduced the tenets of Translational Systems Biology in Chapter 1.1, but after our long trip through the history, thought process, and factors that have led our scientific community to arrive at this place at this time, we will restate what we mean by Translational Systems Biology.

PRIMARY GOAL: FACILITATE THE TRANSLATION OF BASIC BIOMEDICAL RESEARCH TO THE IMPLEMENTATION OF EFFECTIVE CLINICAL THERAPEUTICS

In short, the ultimate goal of biomedical research (as opposed to, say pure biological research) is to enhance human health, and this requires an investigatory strategy that can lead to the engineering of effective therapeutics. The Translational Dilemma represents a breakdown in the process by which pure knowledge is turned into applied knowledge. Translational Systems Biology is aimed specifically at this problem. To accomplish this, Translational Systems Biology was developed using *three Primary Design Principles*:

1. *Utilize dynamic computational modeling to capture mechanism.* Given that our goal is to be able to engineer therapeutics, we are essentially attempting to develop methods of controlling human biology. Methods of

control require an inference of mechanism in the process to be controlled. Mechanism also implies the notion of cause and effect, which necessarily involves recognition that dynamics are important. Translational Systems Biology uses the advances in computing technology to target this requirement.

2. *Develop a framework that allows "useful failure" à la Popper.* The Scientific Cycle is fundamentally iterative; implicit in this process is that there is continued refinement of the knowledge generated. Refinement implies effectively learning from our mistakes. The current Scientific Cycle, as it manifests at the level of clinical evaluation of therapeutics, does a very poor job of providing constructive feedback on the *intellectual* causes of failure; i.e., asking the question "was the target for the potential drug well conceived?" Rather, current feedback focuses almost exclusively on the methodological issues inherent to the execution of a clinical trial (which we have reviewed in Chapter 2.4), obfuscating the process of designing better drugs. Translational Systems Biology aims at bringing transparency to the translational process, and provide for more explicit and directed feedback between experimental results and the conceptual models underlying those experiments. In so doing, this approach strongly reinforces the scientific principle of Falsification as described by Karl Popper.

3. *Ensure the framework is firmly grounded with respect to the history and philosophy of science.* Ultimately, we trust and believe in the Scientific Cycle and the over 2000 years of history and philosophical genius that went into its development, description, and utilization. We believe that the Scientific Cycle is fundamentally sound, that the development of interventions and controls requires the generation and testing of mechanistic hypotheses, and therefore any solution to the current Translational Dilemma must arise from the legacy of the Scientific Cycle. Of course, we might be wrong (which is itself a mindset consistent with Scientific skepticism), which is why Translational Systems Biology emphasizes process over specific knowledge targets; it does not rely upon a particular technology or method, but rather grafts on an ever-growing list of capabilities to meet identified robust and persistent scientific functions.

In order to be effective, design principles need to find tangible expression in order to move from a concept to a practice. For Translational Systems Biology, the implementation of our three articulated Design Principles takes the form of three overall methodological targets and strategies:

1. *Use dynamic computational modeling to accelerate the preclinical Scientific Cycle by enhancing hypothesis testing, which will improve efficiency in developing better drug candidates.* This goal is consistent with our recognition of the current bottleneck in the Scientific Cycle: the step of hypothesis testing. We propose that dynamic computational modeling can allow the visualization and evaluation of mechanistic hypotheses in an accelerated fashion, limited instead by the flow of electrons and photons within computers and the imagination and intuition of researchers, as opposed to the growth rates of cells or experimental animals. Doing so however requires the use and integration of methods applied to the other phases of the Scientific Cycle, i.e., high-throughput experimental platforms for real-world validation, generation of "omics" data sets representing enhanced observational capability, and Big Data analyses to identify new correlations. Dynamic knowledge representation, as an explicit goal of model development that we will discuss shortly, is a critical concept in this process.

2. *Use simulations of clinical implementation via in silico clinical trials and personalized simulations to increase the efficiency of the terminal phase of the therapy development pipeline.* This goal is consistent with our emphasis on improving human health. This focuses the targeted modeling goal on generating populations of simulations that can mirror how the biology manifests at the patient/clinical level, and looks to patient/epidemiological/clinical data as validation/verification metrics. Keeping this goal in mind underlies all aspects of model development within the program of Translational Systems Biology, and helps distinguish this from other approaches which make use of mathematical and computational modeling in biomedical research.

3. *Use the power of abstraction provided by dynamic computational models to identify core, conserved functions and behaviors in order to bind together and bridge between different biological models and individual patients.* This goal harkens back to our recognition that not every detail of a system need be represented in order for an effective control of that system to be designed. This strategy serves three primary purposes. First, it releases us from the pervasive view that "more is better" when it comes to mechanistic detail. Second, it provides a tangible practical target of "sufficient knowledge" as a sequence of models is developed. Third, it provides a formal and rational guide to assess what aspects of mechanistic biology can be considered similar through the range of preclinical and clinical biological systems. This, in turn, will reduce the set of unknown factors that can be invoked to try and explain individual heterogeneity and allow better recognition of what essential biology is conserved from circumstance to circumstance. Intrinsic to this goal is the need to dynamically model states of baseline health from whence disease states arise.

A new field does not exist in a vacuum; there is a preexisting landscape of research into which it must fit. To draw an analogy from ecology, there are multiple niches in the Scientific Cycle for a particular research environment. Many of those niches are filled, and filled well, but they cannot solve what they are not designed to do. Translational Systems Biology is intended to fill the niche represented by the Translational Dilemma. This recognition informed the development of the methodological targets and strategies just listed above: those criteria were designed to fit into the current gaps formed by other fields such as classical Systems Biology, Big Data-oriented Computational Biology, Bioinformatics, Mathematical Biology, and Bioengineering. In addition, in order to further define what Translational Systems Biology is, we think it is useful to also list what Translational Systems Biology is *not*:

1. Translational Systems Biology is *not* using computational modeling solely to gain increasingly detailed information about biological systems.
2. Translational Systems Biology is *not* aiming to reproduce detail as the primary goal of modeling; the level of detail to be pursued needs to be justified from a translational standpoint.
3. Translational Systems Biology is *not* aiming to develop the most quantitatively precise computational model of a preclinical, or subpatient level system.
4. Translational Systems Biology is *not* just the collection and computational analysis of a wide range data types, even if they span a wide range of scales of organization spanning the gene to socioenvironmental factors, in order to provide merely a broad and deep description of a system.

We reiterate that the implementation of a program of Translational Systems Biology does not preclude the ongoing reductionist experimental investigations, the development and utilization of systems biology approaches to quantify fine molecular detail, or progress and utilization of Big Data-oriented computational biology. All these are important, vital aspects within the Scientific Cycle. But, given our recognition that these strategies alone are insufficient to meet the Translational Dilemma, we have crafted the description of Translational Systems Biology to limit overlap with those pursuits, and by so doing emphasize what is missing from all those approaches. Of these other methods, the one most closely approximating Translational System Biology is classical Systems Biology. At one level, this is an obvious linkage, if by the name than nothing else. In fact, Systems Biology has been defined in many ways [1–3], but is generally considered an analytical approach to biologic data aimed at representing "systems level" behavior in various disease states. However, in practice, classical systems biology makes extensive use of highly reductionist experimental platforms, such as yeast, fruit flies, or primitive worms, or focuses on the highly detailed characterization of disease subcomponents, such as molecular profiling of tumors [4–15]. Despite the richness of these approaches, in classical Systems Biology there is still a relative paucity of techniques that transcend into the clinically relevant arena; for this reason, the Translational Dilemma is the "bridge too far" for classical approaches of Systems Biology.

We have developed the concepts of Translational Systems Biology over the course of over 15 years. When we began the line of investigation that eventually resulted in the development of Translational Systems Biology, we had an inkling of what needed to be done, but obviously our early experiences led to refinements and issues that we could not have been anticipated at the outset. To a great degree, this experience informed our concept that Translational Systems Biology needed to be a robust and scalable process, proceeding in some ways agnostic to the specific knowledge and problems being examined. This realization led us to our current emphasis on grounding Translational Systems Biology in the philosophical foundations of science. But, as is so often the case, there was certainly a driving biological question behind this endeavor.

Our research and clinical backgrounds were in the area of acute inflammation and, specifically its manifestation in the diseases of trauma and sepsis. This turned out to be fortuitous not only because of the total lack of approved therapies for these complex syndromes, but also because, as we have previously alluded to, inflammation is almost uniquely positioned as a biological process that intersects with nearly all aspects of human health and disease. This is the case because inflammation, viewed as the means of an organism's ability to deal with injury, can be considered, in addition to growth, movement, and reproduction, an essential quality of living things. As a result, as we delved deeper into the mechanics of inflammation our initial investigations into sepsis and trauma organically grew to involve nearly the entire spectrum of disease. Key to our understanding was the recognition of the Janus-faced nature of inflammation, as a finely tuned, dynamic process required for immune surveillance, optimal repair, and regeneration after injury [16–18], but which, when dysregulated, underlies many complex diseases (e.g., sepsis, infectious disease, trauma, asthma, allergy, autoimmune disorders, transplant rejection, cancer, neurodegenerative diseases, obesity, and atherosclerosis). The paradoxical nature of inflammation required a means of analysis and investigation that could characterize the good with the bad, and allow us to potentially parse the tipping points from one circumstance to the other. It was in this context that Translational Systems Biology came into being.

In the following section, we will discuss the role of dynamic computational modeling in Translational Systems Biology, specifically as a means of *dynamic knowledge representation*. This will be followed by a description of how dynamic knowledge representation through computational modeling can be used to develop a rational framework for the development and implementation of new potential biological therapeutics. This framework will outline how we propose that the tenets of Translational Systems Biology can be used to accelerate the generation and evaluation of promising ideas through the preclinical drug candidate pipeline; increase the efficiency and yield of clinical testing through the performance of *in silico* clinical trials; and, finally, provide a mechanism by which the knowledge generated from population-based evaluations of therapeutic strategies can be utilized and implemented to provide personalized medical care. Then, in Section 4, we will provide specific examples of the tools and methods utilized in Translational Systems Biology, and how they have been able to enhance our understanding of how to approach the clinically relevant study of inflammation.

References

[1] Kitano H. Computational systems biology. Nature 2002;420(6912):206–10.
[2] Kitano H. Systems biology: a brief overview. Science 2002;295(5560):1662–4.
[3] Sauer U, Heinemann M, Zamboni N. Genetics. Getting closer to the whole picture. Science 2007;316(5824):550–1.
[4] Abrams WR, Barber CA, McCann K, Tong G, Chen Z, Mauk MG, et al. Development of a microfluidic device for detection of pathogens in oral samples using upconverting phosphor technology (UPT). Ann NY Acad Sci 2007;1098:375–88.
[5] Ahrens CH, Wagner U, Rehrauer HK, Turker C, Schlapbach R. Current challenges and approaches for the synergistic use of systems biology data in the scientific community. EXS 2007;97:277–307.
[6] Barker PE, Wagner PD, Stein SE, Bunk DM, Srivastava S, Omenn GS. Standards for plasma and serum proteomics in early cancer detection: a needs assessment report from the national institute of standards and technology—National Cancer Institute Standards, Methods, Assays, Reagents and Technologies Workshop, August 18–19, 2005. Clin Chem 2006;52(9):1669–74.
[7] Brownstein BH, Logvinenko T, Lederer JA, Cobb JP, Hubbard WJ, Chaudry IH, et al. Commonality and differences in leukocyte gene expression patterns among three models of inflammation and injury. Physiol Genomics 2006;24(3):298–309.
[8] Calvano SE, Xiao W, Richards DR, Felciano RM, Baker HV, Cho RJ, et al. A network-based analysis of systemic inflammation in humans. Nature 2005;437(7061):1032–7.
[9] Cobb JP, Mindrinos MN, Miller-Graziano C, Calvano SE, Baker HV, Xiao W, et al. Application of genome-wide expression analysis to human health and disease. Proc Natl Acad Sci USA 2005;102(13):4801–6.
[10] Cobb JP, O'Keefe GE. Injury research in the genomic era. Lancet 2004;363(9426):2076–83.
[11] Klass M, Gavrikov V, Drury D, Stewart B, Hunter S, Denson DD, et al. Intravenous mononuclear marrow cells reverse neuropathic pain from experimental mononeuropathy. Anesth Analg 2007;104(4):944–8.
[12] Kourtidis A, Eifert C, Conklin DS. RNAi applications in target validation. Ernst Schering Res Found Workshop 2007(61):1–21.
[13] Liu T, Qian WJ, Gritsenko MA, Xiao W, Moldawer LL, Kaushal A, et al. High dynamic range characterization of the trauma patient plasma proteome. Mol Cell Proteomics 2006;5(10):1899–913.
[14] Steinfath M, Repsilber D, Scholz M, Walther D, Selbig J. Integrated data analysis for genome-wide research. EXS 2007;97:309–29.
[15] Tanke HJ. Genomics and proteomics: the potential role of oral diagnostics. Ann NY Acad Sci 2007;1098:330–4.
[16] Nathan C. Points of control in inflammation. Nature 2002;420(6917):846–52.
[17] Hart J. Inflammation. 2: its role in the healing of chronic wounds. J Wound Care 2002;11(7):245–9.
[18] Hart J. Inflammation. 1: its role in the healing of acute wounds. J Wound Care 2002;11(6):205–9.

3.2

Dynamic Knowledge Representation and the Power of Model Making

What I cannot create, I do not understand. **Richard Feynman**

Far better an approximate answer to the right question, than the exact answer to the wrong question, which can always be made precise. **John Wilder Tukey**

All models are wrong, but some are useful. **George E.P. Box**

The sciences do not try to explain, they hardly even try to interpret, they mainly make models. By a model is meant a mathematical construct which, with the addition of certain verbal interpretations, describes observed phenomena. The justification of such a mathematical construct is solely and precisely that it is expected to work. **John Von Neumann**

What is a model? Often this is a loaded question scientifically, with the answer potentially revealing deep dogmatic divides and embedded prejudices. These cases, in which an unsatisfactory response may be received with a "my way or the highway" attitude and fundamentally determine whether the conversation will continue or reach a dead end, represent highly significant barriers to the development of multidisciplinary collaborations. Obviously, this is not a productive situation, but unfortunately, given the fractured and compartmentalized nature of the scientific community, this scenario is not as infrequent as one would hope. We have dealt with this issue in the process of developing Translational Systems Biology as an intrinsically multidisciplinary endeavor [1]. Here, however, we will focus on addressing that issue in discussing a central method employed in Translational Systems Biology, namely the use of computational modeling as a means of dynamic knowledge representation. In keeping with our procedure throughout this book, we begin our discussion about modeling by providing a logical systematic process by which we will arrive at a definition that we will subsequently use. This process is particularly important when talking about "modeling" or making "models," as multiple different domains mean very specific and different things when they use this term. This is Bacon's Idol of the Market, where ambiguity and disagreement in terminology can disrupt the scientific process, in full force. We will start with the most general definition of a model, and progressively pare away those interpretations that do not apply to our use of models for dynamic knowledge representation.

First, we consider the most general definition of a *model as any representation that is used as a proxy to study a system.* Implicit in this definition is the concept that the model is not exactly like the system being modeled; this allows us to avoid the situation Borges described in *On Exactitude in Science*, which we quoted at the beginning of Chapter 2.3. As such, every model is some lesser or incomplete representation of its referent system. Stated slightly differently, a model is the product of some sort of *abstraction* of the reference system. This abstraction can take many forms, each representing a focus on one particular aspect of the system at the expense of being able to account for others. One simple example is size: a scale model of a building is intended to represent the three-dimensional form of the actual building, only smaller. This model gives a good visualization of what the building will look like, but at the expense of being able to provide other information about the structure, such as being able to calculate the physical/mechanical stresses of the completed building (for instance). Another example is detail: that same scale model of the building will generally not have miniaturized electrical wiring and plumbing, and therefore would not be able to show how the building might function in practice. A third example is an abstraction of form. An example of this type of abstraction is the representation of a system's characteristics or qualities with something other than another physical/tangible object. To keep with our example of a building, such an abstraction could be the plans of the building, or a list of components needed to make the building, or a set of instructions on how to make the building.

The relationship between a model and its referent system is called a *map* or *mapping*. We have already been introduced to this concept in our discussion about the history of science (see Chapter 2.1), and seen how this concept can be derived from the condition of Plato's Cave. In his parable, there is a pejorative implication associated with the differences between the various shadows observed and the "real" object casting the shadow from behind; the implication is that the shadows obfuscate intellectual enlightenment. However, we can also see another, less negative interpretation of this allegory, one that suggests that understanding how such shadows differ can lead to insight into the nature of the "real" object. Such is the case for using models in the scientific process; by understanding the abstractions inherent to the selection and creation of a particular model, the hope is that the use of models can increase our understanding of the system being studied. The key point to recognize here is that there is no "one" model that replicates the referent system comprehensively, and as such there should be a synergistic relationship between the sets of models used to examine a system. Since each model represents a trade-off between the desired target of its abstraction versus the loss of the model's ability to capture other features, it is reasonable to recognize that the use of different models should somehow complement each other by illuminated different aspects of the reference system. Unfortunately, as seemingly obvious as this statement is, we note that a failure to recognize the complementary effect of multiple modeling methods is exactly the barrier to establishing transdisciplinary teams: as long as each specialty area focuses only on what it can represent and places primacy on that aspect to the exclusion of other viewpoints it is extremely difficult, if not impossible, to establish a productive working relationship. This focus is not in and of itself an unreasonable mindset. After all, that preference almost certainly influenced an individual's decision on what field of science to pursue, and was in turn reinforced by being in that particular scientific silo. Therefore, to a great degree, breaking down these barriers requires a process of reminding proponents of different approaches of the actual limitations of their methods. Now, we recognize that this is an extremely difficult and arduous process. Given that the dogmatic dictates and insular environments of professional science significantly promote and reinforce professional success and accomplishment tied to methodology, attempting to constructively deconstruct the underlying assumptions of a particular domain, be it reductionist experimental biology or biostatistics, often results in arguments akin to those about religious dogma. Note that this is Bacon's Idol of the Theater, the stubborn persistent acceptance of an established academic dogma, in full force.

With this understanding of the role of abstraction in the process of modeling, we ask what types of models are typically found in the biomedical arena, and what abstractions are inherent to their use? The most ready example in biology is the use of certain organisms as models for other organisms. We have seen that the traditional and predominant practice of prior to the Molecular Biology Revolution was that of comparative description to categorize living things. This process necessarily involves identifying what is similar and what is different among organisms, and this naturally evolved into the concept that similar organisms could be used to study each other. The most overwhelmingly common application of this approach was the use of experimental animals to model human physiology and disease. This is the foundation of modern experimental biology, and its importance was only accentuated through the Molecular Biology Revolution, which greatly augmented our ability to obtain finer and finer detail in the characterization of living things. To a great degree, when one speaks of basic biomedical research, one is talking about the performance of experiments on biological proxies: cells in tissue culture along with highly manipulated whole organisms.

However, it is critical to remember that these biological proxies are all models, and therefore subject to the same constraints inherent to all models: they incorporate abstracting assumptions and manifest limitations from those abstractions. This recognition is all too often lost when the output of biological experiments is interpreted. We believe this disconnect is due primarily to two properties of biological models. First, and quite paradoxically, the emphasis on the process of engineering a "good" model obfuscates the gulf between the model and the situation it is intending to model since biological models are so heavily engineered in the first place. The second is the fact that since experimental models deal with intact biological units (i.e., cells, organisms), much of what is actually going on is "hidden" from view. To expand on this latter issue, regardless of the degree of detail understood about a biological object, that knowledge is necessarily incomplete (or else you would not be doing the study in the first place!). As a result, some degree of unknown cause and effect will not be accounted for in the engineering of the biological model and its subsequent use in an experiment. What this means is that *all biological experiments are technically phenomenological*, with inferences of mechanism layered into a hypothesis of what might actually be going on. Here, the relationship between biological proxy models and the shadows of Plato's Cave is particularly apt: the interpretation of the phenomenon is mistaken for the actual generative processes themselves because the "shadow" (i.e., the biological proxy) is itself an intact biological system. On one hand, this is semicomforting: the hidden structure of biology provides a foundation of similarity among living things that is the necessary precondition for using a model to learn something new about how biology works. However, it also means that no comprehensive understanding of a biological model is possible, and therefore any cause-and-effect mechanistic hypothesis derived from that model represents a conjecture. We have touched on this concept in our introduction to Translational Systems Biology by stating that the goal of biomedical research is not ontological truth, but rather an understanding sufficient to allow us to develop and engineer interventions. We will return to this topic shortly with a description of what we mean by dynamic knowledge representation.

However, at this point, we take a step back and look at other types of models utilized in biomedical research. After all, we have acknowledged the need to utilize a wide range of abstracting approaches in order to get the best overall picture of the referent system. The experimental models we just discussed can be thought of generally falling into our examples of smaller, reduced, and less-detailed abstractions of human systems, but retaining the form of biological objects. Alternatively, statistical models can be thought of as nonphysical objects used to represent biological systems, with the abstraction of physical form and function to scalar data. This process also has a long and embedded history in biology. After all, what are the descriptive lists used to characterize organisms other than the conversion of those objects to data points? An example of a premathematical "statistical" evaluation is determining that one organism is more closely related to a second organism than a third: the ones that share more data points (or, in Machine Learning parlance, "features") are deemed more similar. The formalization of mathematical statistical methods added incredible power to the ability to characterize similarity, and represents, as we have seen, an essential component of the Scientific Method. We have also seen that these mathematical constructs have grown to a staggering level of sophistication, able to incorporate a multiplicity of data points and of coming up with novel descriptions of "similarity."

But, as we have already mentioned in our discussion about the pitfalls of the current biomedical research environment, this is a potential trap if one forgets the role that identification of correlative patterns plays within the Scientific Cycle. Recall that the original intent of statistics was to evaluate the differences or lack thereof between two sets of data representing the differential outcomes from an experiment. Intrinsic to this role is the fact that an experiment existed, one that presumably resulted from a cognitive and deliberate construction of a testable hypothesis. As such, there existed a defined intervention, chosen specifically in order to test the hypothesis. Alternatively, if you now have a series of sets of data, say, "existing in the wild" prior to the performance of an experiment, you can use these same statistical measures to try to parse those sets of data to say: "This set has more in common with that set." This recognition can then be used to formulate a hypothesis that might account for those similarities. Fit within the Scientific Cycle, this step involves contextualization of the acquired new information into the sum of preexisting knowledge, and coming up with a best "guess" (i.e., Abduction à la Peirce) about how the new observations came about. The result here is a new hypothesis that, given the structure of the Scientific Cycle, requires an experiment to test. But, at this point in the process it should be immediately obvious that applying another correlative test to the original data set, absent a new intervention, cannot be used to test that hypothesis. Rather, *doing so merely provides another aspect view of the original system!* Furthermore, the ability of that statistical model to project any statements about that original data set into the future, i.e., *prediction*, is completely dependent upon the assumption that the essential dynamics and generative structure of the overall system will persist into the future sample period. For example, the success Nate Silver had in being able to predict the outcome of the 2010 United States Presidential Election, seemingly uncanny as it was, was possible primarily because the fundamentals of the US electorate, in terms of composition and especially geographical distribution, remained essentially unchanged from the prior elections used as a training set for his model. The result is, while this powerful, predictive correlative model can tell you what is likely to happen, it cannot tell you what to do if you want to change the outcome. *In short, the model can tell you if you are going to lose, but it cannot tell you what to do in order to win.* Doing so requires an analysis of the data in the context of prior and existing knowledge, and an inference of mechanistic process that led the result. But, if you do all that, then all you have at the end of this analysis is a hypothesis that needs to be tested. This, of course, is a very difficult thing, since Presidential elections do not avail themselves to experiment. In the biomedical arena, this is the difference between prognosis/diagnosis versus therapy. Working in the biomedical domain, we at least have the opportunity to perform and design experiments to potentially test those generated hypotheses. However, as we have seen in Chapter 2.4, there are significant hurdles and constraints on the performance of the terminal level of experiment we need to perform, i.e., the Clinical Trial.

For prediction outside the range of conserved dynamics that generate the training data set, we must turn to another type of abstraction, one that is also an abstraction of form and one that utilized mathematics, but now emphasizing the existence of a mechanistic universe. These are dynamic or mechanistic mathematical models; *dynamic* in the sense that they inherently involve characterizing the system as it passes through time, and *mechanistic* in so far they incorporate inferences as to how that passage of time is manifested by the system. At this point, we think it is important to distinguish between a *mechanistic model* versus the common perception of "mechanistic" representation in biology. Perhaps the easiest way to think about this distinction is that for a mathematical model, *mechanism* is equivalent to a *rule*; i.e., the mathematical model provides for a rule that accounts for how the system evolves over time. For example, Newton's equations of motion are considered both dynamic and mechanistic: they represented the behavior of bodies in motion over time based on the forms of their equations. We have since learned of the deeper quantum and relativistic "mechanisms" underlying the phenomena reproduced by Newtonian Mechanics, but that information does not obviate the general utility of the rules expressed in Newton's equations. So, too, in biology, where mathematical models of cardiac and pulmonary physiology are still utilized effectively today, even though we know that the generative biology underlying cardiac output, blood pressure, and tidal volume lie multiple levels of organization below.

Note that the construction of a dynamic mechanistic mathematical model requires a hypothesis at the outset; this is opposed to a statistical/correlative model, in which a hypothesis is the output resulting from the model. As such, the complementary nature of dynamic mechanistic models and correlative statistical models within the Scientific Cycle is evident immediately. Looking at the standard of "prediction," the difference is also quite clear: the predictive capability of a dynamic mechanistic model extends beyond the model's ability to "replay" an existing system, instead it allows one to say what will happen if you change that system in a specific way. A trusted predictive dynamic mechanistic model can thus ascend to the level of a *theory*, a rule powerful enough in its domain of application capable not only of telling you what will happen, but also of telling you what absolutely *cannot* happen. The generation of such theories is the ultimate goal of pure science, and it is in the service of this goal that the iterative Scientific Cycle exists. So too, in play, are the cautions of Hume, Popper, and Peirce: science demands that we be aware of the limits of any hypothesis we would try to elevate to the level of theory. To a great degree, biomedical science can avoid this standard: we merely want to know enough to be able to exercise reasonably trustworthy control. But the goal is still to bind together the heterogeneous observations with underlying mechanisms that allow us to explain how those divergent behaviors arise; this is the key goal that would actually allow us to accomplish the translational goals of moving knowledge across biological proxy platforms, to patient populations and then between individual patients. In order to accomplish this goal of establishing trust in our knowledge (as expressed through models), the fundamentals of the Scientific Cycle remain.

At this point, it is valid to ask about the costs and pitfalls of dynamic mechanistic mathematical models. Here, we return to the mapping problem, and perhaps an unfortunate legacy of Physics Envy. The elegance of Newtonian Physics and the Calculus is the perfect matching of tool and application. This elegance is reinforced by the relative simplicity of the forms those rules take; few human ideas are more elegant than $F = ma$. Moreover, $F = ma$ can be taught to middle school children. The degree of mapping did not occur by accident: Newton developed them both (with all due credit also to Leibnitz on the Calculus front) as necessary means to an end. But the power of Newtonian Mechanics often causes us to forget some of the basic assumptions he made, ones that clearly are not true in the "real" world: massless points, frictionless surfaces, and pure Euclidian geometry. The power of Newtonian Theory is such that, for many uses, these unrealistic assumptions do not matter. However, biology does not have a correspondingly powerful and well-mapped mathematical representation (yet!). Attempts to map biological systems onto the mathematics of physics result in sets of abstractions and assumptions that often readily break the resulting models. Such examples are mathematical models that assume idealized growth rates, or uniform and consistent distributions of mediators, or idealized geometries. When faced with the "messiness" of biological systems, such types of abstract idealizations cause a near immediate breakdown of trust on the part of the biologists. What is needed, then, is a way to more readily integrate the use of dynamic mechanistic mathematical modeling into the biomedical arena, doing so in a way that can more closely match how biologists are accustomed to thinking about their systems.

This now brings us back to summarize the types of models used in biomedical research. We first have biological proxy models used in experimental biology. These models have the benefit of being "real," but at the expense of having perpetually hidden processes. Next, we have statistical models, which are used both to evaluate the results of the proxy model experiments and to facilitate the generation of hypotheses. These models serve an integral role in the interpretation of experiments, and to some degree facilitate the process of sorting through, and finding potential meaning in, the increasingly massive data sets now available. However, we have seen that statistical models are limited by their intrinsic correlative nature. Finally, we have dynamic mechanistic mathematical models, which represent mathematical expressions of biological rules. While incredibly powerful, these models need to be referred back to the "real world" in order to ensure that they have not traveled too far afield in making the abstractions that go into the particular model. For a biologist, this means not making assumptions that are, in their minds, "clearly wrong." This turns out to be quite a difficult task, since as we have seen the *modus operandi* of biology has been to emphasize detailed representation of particular systems of interest. It is therefore extremely difficult, if not nearly impossible, to represent a biological system to the degree of comprehensive detail that will completely satisfy a biological researcher: any mechanism-based model will always be subjected to the charge that "you left something out." How can we address this issue?

We suggest that the solution to this challenge involves mapping not only to the specific knowledge present about a particular system, but also to the process by which biologists acquire, interpret, and process their knowledge. Key to this process is the recognition that biologists deal with incomplete knowledge all the time; as we discussed earlier, the fact is that the nature of biological proxy models is that there will perpetually be "hidden" functions involved in their response within the experiments they are used for. This limitation means that despite the most controlled and reductionist conditions, and regardless of how closely a biological proxy model matches the researcher's conception of how the referent system should work, there is always going to be the possibility and even likelihood that there are other factors in play that could be driving the behavior of the model. Biological researchers understand this limitation, of course. This is why "additional mechanisms" are nearly always invoked when an experiment provides unexpected results. Actually, "additional mechanisms" are often also invoked during discussion about successful experiments, so much so that the phrase: "The results of this paper do not preclude other factors/pathways/genes also being involved in the behavior of the system," can be found in nearly every molecular biology publication. In fact, the ubiquity of this statement and the underlying realization that it represents renders the statement nearly meaningless, because the research community simply does not have a general strategy for handling the implications of such a statement. That the standard response strategy is the disclaimer that "Additional studies are required," misses completely the implication of the original recognition: knowledge will *always* be incomplete, and therefore, additional studies can *always* be called for.

This is what the physicist Robert Rosen, one of the founders of Theoretical Biology, meant when he talked about the dangers of "infinite regress" in his book *What Is Life?* Within this scientific paradigm, there is no end point. On one hand, this is not the end of the world if we are talking about pure science; in actuality, this is kind of the point of pure science, the perpetual quest for more information and knowledge, seeking the next horizon. It is a romantic notion, stirring the explorer within us all. But when we are operating in an applied science like biomedical research, this standard cannot be held: there must be some point where incomplete knowledge is *sufficient* to allow for the development of control strategies. From a practical standpoint, researchers in the biomedical arena are aware of this; if it were not so, then no pharmaceutical company would ever start development of a potential drug.

The issue is not that the need to identify sufficient knowledge is recognized; rather, we would suggest that the *means* of defining sufficient knowledge is not appropriately scientifically rigorous. The problem is that the current standard relies upon interpretation of the same set of biological proxy models that, if questioned about with enough vigor, nearly all biological researchers would acknowledge represent only shadows of the real system/process being targeted. The pragmatic modifier of such an admission, however, is that there is not anything better. But we suggest that there is.

We would assert that the solution to this process is developing an assessment of whether or not the hypothesized mechanism, incomplete as it is, is actually internally consistent and behaves as the researcher perceives it. Mechanism-based computational/mathematical models can be seen as a means of *dynamic knowledge representation* to form a basis for formal means of testing, evaluating, and comparing what is currently known within the research community. In order to be able to "see" the consequences of a particular hypothesis-structure/conceptual model, the formally represented knowledge is moved from a static view of relationships (as depicted in a flowchart or state diagram) to a dynamic model, one in which the mechanistic consequences of each hypothesis can be observed and evaluated. We term this process *conceptual model verification*: dynamic representation of a conceptual model as a means of its *verification*, analogous to model checking in computer science, i.e., determining whether the computational model performs as expected based on its construction. Conceptual model verification has the effect of providing a logical check on any hypothesis that a researcher may come up with.

This dynamic knowledge representation does not question as to why the researcher believes what she believes; those questions are answerable by examining the other steps of the Scientific Cycle involving data collection, determination of correlation, and construction of an abduced mechanism. The dynamic knowledge representation also does not include any "hidden" functions. Rather, it is a bare-bones expression of the hypothesis being posed. As such, a dynamic computational model that instantiates a mechanistic hypothesis will behave exactly as it is constructed, incorporating only the relationships inferred within the particular hypothesis: no more, no less. Once constructed, this computational expression of hypothesis can now be treated as an experimental object, just as a biological proxy model would be: to be manipulated experimentally and its behavior compared against available real-world data and observations. The computational model will either behave sufficiently similarly to these real-world data (by whatever criteria are chosen), or will not. Importantly, even if the model matches real data sufficiently well, it does not mean that the computational model and the knowledge it represents are "true." This finding just means that the model is "sufficient" in reproducing observable behavior. On the other hand, if the dynamic computational model does not match the real-world observations, then this means that the model, admittedly incomplete as it is, is insufficient, and then more detail needs to be added. This latter condition is actually the critical one, as it plays a critical role in the falsification goal for science as posed by Popper.

By utilizing dynamic knowledge representation through dynamic computational models, researchers can potentially greatly accelerate the process of experimental testing of their hypotheses, since now this is being done *in silico* as opposed to through biological proxies. Viewed in this way from a scientific process standpoint, the use of dynamic knowledge representation is akin to performing thought experiments, those mental exercises so valued in physics. Now, however, the ability to instantiate these thought experiments through computer modeling provides tangible outputs that can be fed back into the Scientific Cycle. The key point here is that the utilization of dynamic knowledge representation in this fashion merely mimics what every good researcher already does, but more explicitly and formally. This process paves the way for "useful failure" by replaying that grade school math trope: "show your work."

Note also that this concept of using mechanism-based computational/mathematical models for dynamic knowledge representation substantially shifts the goal of model creation. Now, rather than attempting to develop the most detailed model that reproduces all the factors and features present in the system (often the target of systems biology), the goal now is to develop minimally sufficient expressions of hypotheses that fits a set of reference data. As such, it is less important to have quantitatively predictive models that target some limited aspect of biology at this point than to have qualitatively plausible models that represent more systemic and global phenomena. To restate the quote from statistician John Tukey at the front of the chapter: "Far better an approximate answer to the right question, than the exact answer to the wrong question, which can always be made precise." This mindset shifts the burden of proof for what constitutes a "useful wrong model" (to paraphrase the statistician G.E.P. Box) substantially, as more time and effort can be spent cycling through simulated thought experiments using dynamic knowledge representation than attempting to fit a model quantitatively to some high degree of precision. This approach treats dynamic computational models as disposable objects, targeted for falsification and limited, but well defined, ranges of usefulness.

The development and use of mechanism-based computational/mathematical models carries with it an additional benefit: the need to explicitly express and implement one's hypothesis structure is itself an extremely beneficial process. A researcher may be as diligent as possible, but there is still a tangible benefit to the process of constructing a model [2]. The physicist Richard Feyman notably stated: "What I cannot create, I do not understand;" to paraphrase him, we would restate that "to construct is to gain understanding." This is actually a tenet of the Constructivist philosophy of education developed by Jean Piaget; while initially intended for primary education, the potential application of this philosophy through dynamic knowledge representation in biomedical research demonstrates that we never stop needing to learn.

This concept of dynamic knowledge representation and its implementation through the construction of mechanism-based dynamic computational and mathematical models is central to the concept of Translational Systems Biology, and will be described in more depth in Chapters 4.3–4.5, where specific examples of various types of computational modeling in the Translational Systems Biology of inflammation will be presented. This concept also plays an integral role in achieving the other central tenets of Translational Systems Biology, as will be discussed in the next several chapters.

References

[1] An G, Hunt CA, Clermont G, Neugebauer E, Vodovotz Y. Challenges and rewards on the road to translational systems biology in acute illness: four case reports from interdisciplinary teams. J Crit Care 2007;22(2):169–75.

[2] An G, Nieman G, Vodovotz Y. Computational and systems biology in trauma and sepsis: current state and future perspectives. Int J Burns Trauma 2012;2:1–10.

3.3

A Roadmap for a Rational Future: A Systematic Path for the Design and Implementation of New Therapeutics

By now we believe that we have made the case that the nature of the primary problem facing the biomedical research community today is the Translational Dilemma, which is manifest in the barriers to translating basic science knowledge into effective clinical interventions. We have also examined, from a historical, philosophical, and social standpoint, the origins of this situation. We are clearly not alone in having pointed out some of the same failures of the drug pipeline; one very interesting and condensed view can be found here: http://www.randombio.com/drug-failures.html. Our proposed approach, Translational Systems Biology, is designed to address the various aspects of the translational challenge of being able to effectively engineer therapeutic interventions for complex, systems-level diseases. In the pursuit of that goal, we have proposed a scientific methodology—dynamic knowledge representation based on dynamic computational models—which can be employed in the service of achieving the stated goals of Translational Systems Biology. Now, we will outline how we see that method be applied to the various stages of the drug development and implementation process, with the specific purpose of increasing the pipeline's efficiency and yield. Because the ultimate goal of whole biomedical research endeavor eventually converges with a doctor telling her patient what is recommended for their particular situation, our focused goal is to enhance human health by facilitating the generation of better drugs, or, perhaps more correctly, better drug-based strategies to meet the needs of individual patients. We assert that many of the current dilemmas involved in picking the appropriate regimen for a particular patient can be traced back to the fact that the actual tools in our therapeutic toolbox just aren't so good. Too often, there must be a trade-off between efficacy and side effects; too often, there is too much uncertainty as to whether a particular patient's biology is appropriately targeted using the drugs that are available; all too often, the best beneficial effects of the available drugs simply do not make enough of a change to the patient's pathophysiology to make a difference.

If our tools were better, these decisions would be much easier. This sounds deceptively simple, but of course the devil is always in the details. What we propose is a framework for engineering the process of engineering and delivering new therapeutics. We have already identified the limitations of the current biomedical research landscape that have led to this situation, and categorized those issues into various phases within the drug development pipeline: the identification and evaluation of promising drug candidates at the basic science/preclinical level, the testing of potential therapeutics at the clinical level, and the implementation of such therapeutics for individual patients. Below we introduce how Translational Systems Biology can be used to address the deficiencies and challenges present in each of these phases.

RATIONAL EVALUATION OF DRUG CANDIDATES: KNOWING WHAT MIGHT WORK, AND MORE IMPORTANTLY, WHAT WON'T

In this section, we will discuss some concepts for Translational Systems Biology as applied to the preclinical drug candidate evaluation process. Note that we are not talking only about rational drug *candidate discovery*: this is a field in of itself, and a valuable one. There has been some improvement in this task through the adoption of "omics" methodologies, predicated on the idea that if one can characterize, at a gene/protein/metabolic level, the responses of cells and tissues in various pathophysiological conditions, this will lead to increased understanding of those diseases. This increased understanding will, in turn help in identifying molecular targets for potential drugs. Furthermore, knowing something about the molecular structure and function of these potential targets can aid in the biochemical design of molecular compounds to modulate the target's function. However, while we think these are very important on the path to rational drug development, we also believe that a focus on this area does not address what we view as the primary source of failure in the current drug development pipeline. Despite the advances in the multiple high-throughput bioinformatics strategies to find the specific docking regions of target molecules, or the effect of drug binding on subsequent protein function, or how genetic predisposition might be correlated to existing panels of drug candidates, there is increasing recognition that more data does not necessarily lead to better drug targets. Thus, these "omics" methods have not proven to be the panacea for the design of drugs, clinical trials, and diagnostics that they were projected to become [1–3].

The mantra of the pharmaceutical/biotechnology industry has been "fail early, fail fast, fail often." However, not all failures are the same, and this distinction is what concerns us. Rather, we would say (and we think that the pharmaceutical industry would agree) that the main problem is the overwhelming attrition seen in "promising" candidates as they make their way through the development pipeline. The modes of candidate failure, ranging from lack of sufficient efficacy to unacceptable side effects, represent failures of the conceptual models about how these drugs are supposed to work. What is missing in the current drug development pipeline is a means of determining and evaluating whether the system-level consequences of attempting to intervene in a particular pathway will lead to a beneficial outcome, balancing safety versus efficacy. In short, we seek a means to answer the question: "Is targeting this particular pathway a good idea?"

This is not to say that there has not been some movement in this direction. For instance, the classical systems biology community has utilized mathematical modeling and simulation to study a range of subcellular and cellular processes [4,5], providing high-resolution characterization of multiple signaling pathways and gene regulatory networks. The depiction of specific, key regulatory pathways has thus become a major area of focus for pharmaceutical industry as they attempt to better understand the targets of their future interventions [6,7]. However, while this type of mechanistic computational modeling that "ends at the cell membrane" is inherently useful to an industry focused on screening for drugs that modulate specific molecular pathways, the pathway-centric process is inherently dissociated from later translational steps in the drug development process as potential effects are scaled to multiple tissues and organ systems of intact organisms. A given drug candidate must not only be identified in the most efficient way possible; this compound must also be passed through toxicity testing and clinical safety and efficacy studies. While the mechanism of action of a given compound at the molecular/cellular level may, through mathematical modeling, be predictable with exquisite precision in highly controlled laboratory experiments (as is the paradigm of classical systems biology), the effects of such a compound *in vivo* are in no way predictable from such models. Given the multiscale complexity of the disease processes we seek to control, it is virtually impossible to make reliable inferences as to the reconstituted consequences of intervening at one selected point. While it is impossible to anticipate the unknown, we propose that it would be incredibly beneficial to be able to assess whether or not a particular hypothesis structure or conceptual mechanistic model is internally consistent and behaves as expected.

This is why we have introduced the concept of *conceptual model verification* to identify and clarify the dynamic consequences of a particular conceptual model, a procedure by which static diagrammatic representations of conceptual models are brought "to life." This is the impetus for our introduction and use of *dynamic knowledge representation* as a means of instantiating "thought experiments" such that the dynamic consequences of a particular hypothesis structure can be seen [8,9]. As we have seen in Chapter 3.2, this process allows a researcher to "check" his/her preconceived notions, to see if the particular mechanistic hypothesis actually behaves as expected. Dynamic knowledge representation has the potential to point out so-called "unanticipated" effects, which are in fact not unanticipated but rather ignored or underestimated effects of a given perturbation of a complex biological system. Dynamic knowledge representation may be augmented with insights derived from high-throughput and high-content data [9], along with appropriate data analysis and data-driven modeling [10–12], to populate and calibrate dynamic computational models of a relevant disease, individual patient [12], or patient population [13–15].

This positioning of dynamic knowledge representation is completely consistent with the Baconian Scientific Method. Moreover, it can be seen as a natural outgrowth of the history and legacy of scientific practice. Therefore, rather than being a "new kind of science," the methods of Translational Systems Biology are consistent with and incorporate the philosphical and logical underpinnings of science that have been developed in the course of Western European history and proven themselves over the centuries. Presenting Translational Systems Biology in this fashion may not have the impact of the "shock of the new," as would be the case if we were to cast our scientific approach as something drastically different than what has come before (e.g., claiming an "end to hypothesis-driven science," or some other such nonsense). While pronouncements that adopt a "paradigm-shifting" tone and strategy can often lead to a great deal of public enthusiasism and interest, we should realize that such claims are often ultimately hollow. Why? Because... Science... Thus, alternatively, we have chosen to follow the more mundane path that merely builds upon the intellectual legacy of Western Civilization, while updating this approach to the data-rich environment of modern science.

This approach includes recognizing the power of "theory" to bind together a wide range of heterogeneous observations, with Newtonian Mechanics as the prototypical example. In the previous chapter, we discussed both the positive and negative aspects of the Newtonian Physics legacy, but here we emphasize the positive effect of the Newtonian paradigm in providing an example for using mechanistic abstractions as a means of bridging biological system to biological system. The key here is to recognize that this linking of biology to biology is *only possible if abstraction is utilized*; without it, we will find ourselves regressing into the traditional "everything is different and unique" viewpoint of biology. Rather, the use of mechanistic models of appropriate abstraction fits with our intuitive notion that there are "some" similarities between biological organisms, enough so that lessons and properties learned from one biological system may be translated to another. As we have discussed, this is the exact role that dynamic knowledge representation plays, only now transposing an intuitive concept into a formalized expression that can be interrogated and experimented with. *In fact, we propose that the ability of a common mechanistic model to represent the behavior across different biological systems/organisms/individuals actually be the definition of what can be considered functionally similar between those systems.* Viewed in this fashion, different biological systems would represent different parameterizations of a common mechanistic model, a categorization paradigm that both captures the richness of biological diversity as well as its striking similarity. This is a central concept in Translational Systems Biology and its use of dynamic knowledge representation to traverse the course of preclinical research, and we will see it reemphasized in both the performance of *in silico* clinical trials and the personalization of medical therapy. In the context of rational drug discovery in, for example, the setting of inflammatory diseases, this implies that the search for a drug candidate in a new disease can be facilitated greatly by the results of modeling the previous disease. It also means that the results of this disease modeling at the level of drug discovery can be translated relatively easily to modeling the individual or a cohort of individuals in the context of a clinical trial.

Approached in this fashion, Translational Systems Biology involves accelerating the proven Scientific Cycle using dynamic mathematical modeling based on mechanistic information generated in early stage and preclinical research to specifically simulate higher-level behaviors at the organ and organism level and the translation of experimental data to the level of clinically relevant phenomena. The utilization of computational modeling of experimental data in this fashion will facilitate a rapid, parallelized clinical translation of drug candidates, vastly increasing the throughput of candidate evaluation as compared to the current serial process [15]. In this schema, the current linear, time-consuming, expensive, and failure-fraught system by which drug candidates are tested sequentially using *in vitro* platforms, in preclinical biological proxy models, and subsequently through phased clinical studies would be replaced with a workflow in which at every level of drug development a "reality check" against the unifying mechanistic models is carried out. We believe that these operational and conceptual changes derived from Translational Systems Biology will help deliver therapeutic candidates with a substantially greater chance of success to the end of the preclinical portion of the development pipeline. We will see in Section 4, some specific examples of these benefits using a series of different computational modeling methods. But given our clinical focus, we realize that we cannot move forward without specifically addressing that "bridge too far" that has bedeviled the biomedical research community, the Clinical Trial. We will discuss our strategy for crossing this bridge, the use of *in silico* clinical trials, in the following section.

IN SILICO CLINICAL TRIALS: CROSSING THE "BRIDGE TOO FAR"

By now it should be clear that the question that spurred us to embark on our Translational Systems Biology work, was: "why are there no drugs for acute inflammation and critical illness?" [15]. In attempting to answer this question, we have discussed the factors that we consider to be central to this fundamental reality, starting from misconceptions and missteps at the level of basic research, and culminating in deficiencies in the drug development and clinical trial processes. From the outset, our approach to the question posed above was characterized by an assessment of the practical issues involved and a search for practical solutions to crossing this last bridge before a drug could be deemed "successful." Thus, we focused our initial efforts precisely on addressing, with dynamic computational modeling, the issues that arose in the failed clinical trials for acute inflammation and critical illness [16]. To a great degree, our emphasis on the clinical endgame is what led us to flip the standard investigatory script in our development of Translational Systems Biology, with its emphasis on recreating and examining clinically relevant situations. Had we set out with a more traditional, reductionist mindset, we could have taken the approach that we needed to first model molecular interactions within a cell, then cell–cell interactions with a tissue, then tissue–tissue interactions within an organ, then organ–organ interactions within an individual, and finally interindividual variability, to finally arrive at being able to simulate the impact of a drug on a population (i.e., simulating a clinical trial). Instead—and this speaks to the entire rationale behind Translational Systems Biology—we began our efforts where the main problem appeared to be: at the level of the clinical trial. In this chapter, we will cover the rationale for, and progress in, the use of computational modeling to streamline and enhance clinical trials. In Chapters 4.3 and 4.4, we will cover the methodology that underlies these mechanistic computational models in detail; our goal in the present chapter is to focus on the rationale, key results, and lessons learned.

We have set the stage for our discussion about *in silico* clinical trials in Chapter 2.4, with our discussion of the history and rationale of Randomized Prospective Clinical Trials, with particular attention to the many challenges and flaws in the current state of clinical trial design, execution, and analysis. To summarize and recapitulate, the major issues concerning clinical trials can be categorized thusly:

Issue #1. Unrecognized, or unmanageable patient heterogeneity within the trial. These factors include insufficient randomization, poor effect signal to background noise in the trial (inadequate treatment effect), and therapies inappropriately targeted to underlying biological processes.

Issue #2. Inability to gain sufficient statistical power within a trial. These factors include logistical and economic constraints on including/collecting enough patient data, patient enrollment, and the statistical manifestation of issues seen in #1 above.

Issue #3. Confounding operational factors associated with carrying out the trial, and extrapolation of the trial results into the real world. These issues are slightly distinct from the trial enrollment issues seen in #2, and concern the human factors involved in implementing the proposed therapy within the confines of the trial. As such, these issues are related to patient compliance, retention, and the general treatment effect of participating in a trial (i.e., a higher degree of attention than in the general population).

Let us address the last issue first, for the simple reason that Issue #3 is fixed constraint based on how the world operates (i.e., it is messily complicated). To a great degree, these issues, consisting of the practical factors that always intervene with human beings are asked to do something in the course of daily life, represent challenges for psychosocial engineering, and are well beyond the scope of this book. However, we do believe that if we could come up with better solutions to Issues #1 and #2 above we would help considerably those attempts to try and do the policy-practice development that would address Issue #3. For example, the issue of compliance with a particular therapeutic regimen would be significantly enhanced if patients had to take fewer pills, less often. Similarly, if we had a better ability to tell patients with greater precision what was actually wrong with them, it would make deciding what to do for them much less complicated overall. Rather, given the focus of our book and the rationale of Translational Systems Biology on establishing a foundation upon which additional complexity can be added so that we can more effectively implement acquired knowledge at the clinical level, we consider the factors in Issue #3 as confounding our ability to interpret the effects of the proposed therapies being tested at the clinical level. To restate this a bit, the foremost goal in a clinical trial is to see if the proposed intervention *actually works* from a biological standpoint, and it is only after this is determined can any thoughts about how that might be implemented come into play; after all, it does no good to try and figure out how to optimize delivery of a drug that actually doesn't work. Therefore, our goal is to determine whether a proposed biological intervention, having passed through a preclinical development pipeline reliant upon reductionist biological proxy experiments, can actually function as expected in a clinical setting, given the much greater complexity of human disease. In actuality, this is functionally just another step in the iterative use of dynamic knowledge representation for conceptual model verification we have extensively discussed, however the importance of the leap to the human clinical situation warrants specific and particular focus.

Issue #1 can be characterized essentially as the need to have a better idea of what is actually going on within the clinical patient population by somehow capturing the messy heterogeneity seen in the clinical setting. We have already mentioned the increasing desire and practice in clinical medicine of using genomic screening and profiling to increase the level of description possible for the patient population. However, as we hope we have been able to communicate in earlier portions of this book, this type of patient/disease characterization is merely a continuation of the legacy of descriptive list-making in biology. While it might be useful to describe things in order to separate one group from another, it is insufficient to be able to characterize what separates one group from another functionally, or if such a difference is present at all. It is already well recognized, even in the genomics field itself, that the functional consequences of the vast array of genomic differences noted within a population are extremely poorly and sparsely known; this is a central source of uncertainty with regard to the interpretation of the results of such a data output. Alternatively, we assert that patient groups are instead separated by their biological function (almost by definition, since that is actually what we see at the bedside!), and that the depiction and characterization of that functional difference leads us down the road of needing to depict mechanism, and then to the central role played by the hypotheses about those mechanisms. Furthermore, as we have established previously, the assessment of mechanistic hypotheses requires dynamic knowledge representation operating within the iterative Scientific Cycle, to be repeated until those hypotheses attain a sufficient level of trustworthiness that allows them to be targets for the engineering of control strategies. Therefore, the use of Translational Systems Biology in performing *in silico* clinical trials rests upon the basic tenet of using mechanistic representations of function as both a means to describe different groups and in order to evaluate defined interventions. We will see specific examples of how different types of computational modeling approaches are used to accomplish these tasks in Section 4, but as a preview of those discussions we list here the specific benefits associated with the use of these methods in the performance of *in silico* clinical trials:

1. *Enhancement of study group substratification*: In current practice, this refers to the data-driven stratification of patient subgroups, with the idea that individualizing treatment for a subgroup of patients is likely to be better than administration of "one size fits all" therapy to all patients with a given disease, and more realistically achievable than the laudable goal of individual-specific medicine [17]. However, we have already discussed the limitations of the purely descriptive approach. Consistent with our concept of binding together biological observations with mechanistic models, Clermont et al. [18] demonstrated this concept as applied to the creation of simulated patient populations using slight variations of parameter values of an ordinary differential equation model of systemic inflammation. This study (expanded upon in Chapter 4.3) shows how the use of an *in silico* trial to enhance subgroup stratification and candidate patient identification. This simulated patient population provides a higher resolution, more detailed representation of each simulated patient in terms of the response trajectories of inflammatory mediators. This level of description and patient tracking allows us to evaluate closely how they respond to their underlying proinflammatory stimulus, with and without a proposed intervention. This, in turn, allows for the identification of potential biomarker-defined inclusion criteria for a clinical trial. In essence, utilized in this fashion, each simulated patient acts as its own control with respect to the proposed intervention. This type of analysis is functionally impossible to obtain in real-world clinical trial cohorts. Thus, *in silico* clinical trials provide biomedical researchers the ability to play "what if?" scenarios specific to simulated patients with defined inflammatory response patterns. Furthermore, this capability allows the generation of specific types of simulated study populations, for whom social or ethical factors may limit their possible representation in a clinical trial, such as African-Americans (known to be generally underrepresented in many clinical trials), or women of childbearing age (excluded for potential teratogenic risk). We have noted that this is a current significant limitation of clinical trials today, as they are very prone to miss important (positive or negative) effects in subgroups that are sampled inadequately. This missampling can lead to later discovery of adverse events following a promising clinical trial, or, in the failure of truly useful treatments in clinical trials that were not properly targeted to the patients that would most benefit from them. By simulating massive virtual cohorts sampled from the space of potential patients, *in silico* clinical trials can achieve much more thorough sampling of possible patients. The acquisition and analysis of this simulation-generated data can in turn reveal clinical patient subgroups that merit particular attention, and lead to better-informed patient selection criteria and more effective clinical trials.

2. *Augmentation and optimization of protocol design*: Clinical trials represent experiments, and as all experiments they require good design to maximize the likelihood that useful knowledge can be acquired from their execution. As such, protocols for interventions depend on multiple complex and often interacting parameters (e.g., dosage levels, timing and frequency of administration) aimed at giving the proposed therapy its "best chance" of being deemed effective. However, attempting to determine these parameters experimentally over a wide range (and currently poorly functionally characterized) of individuals is impossible from a practical standpoint at present. Rather, investigators "take their best guess" based on extrapolations from the preclinical experiments and preliminary clinical-safety trials that have come before; the limitations of this approach have led us to the Translational Dilemma we face today where promising drug candidates fail at the terminal point of clinical evaluation. The inability to anticipate and account for this degree of interindividual heterogeneity

will doom a clinical trial to failure at the outset. Since they can be used to simulate multiple sets of trial parameters and protocols, *in silico* trials bring in an engineering aspect to the design of clinical trials that was not previously present. Now, as opposed to intuitive impressions or extrapolated statistics, mechanistic models used to generate simulated patient cohorts allow for a more rigorous computational optimization of these parameters, both on massive populations and for individual patients, and will increase the precision with which protocols can be designed, and therapeutic endpoints defined. This approach mirrors (finally) the application of engineering principles seen in practice in virtually every other aspect of human technical endeavor, and moves the field of biomedical research away from a reliance on ad hoc prototyping (i.e., using proxy models) to actual design (i.e., using generative principles).

3. *Enhanced characterization of the control group*: Clinical trials rely on control groups against which the effect of a proposed intervention is compared. However, given the vagaries of clinical practice as well as the fundamental heterogeneity of biology, many control groups may actually compare poorly to the intervention group. As we have noted, interindividual variability in both underlying biology and clinical practice leads to a situation in which the current means of defining of "similarity" between control and intervention patients is often quite crude and imprecise, confounding the ability to actually define the effect of the proposed intervention. *In silico* trials, however, offer the ideal control group: each simulated patient can be simulated with and without the intervention. Comparison of results against these "perfect" controls thus removes a source of uncertainty that is unavoidable in real trials.

4. *Capturing heterogeneity with stochasticity (at first!)*: We have noted repeatedly the varied forms and trajectories seen in biological systems, most notably here in the clinical setting. Ideally, all that heterogeneity could be explained by finding a common generative mechanism (i.e., a theory) and explaining the different individual manifestations through variations in parameters and initial conditions. This is the ideal Newtonian goal that we *should* recognize as being akin to Plato's Ideal Forms, i.e., something to strive toward while realizing that it is something we will likely never attain. However, as we have also argued, knowing that the ultimate ideal goal may be out of reach does not mean we should not utilize all our capabilities to improve our insight; in fact, we should specifically approach such a task while being cognizant of finding "stopping points" in our investigations where our levels of abstraction and representation are sufficiently useful. This, to a great degree, is what is driving us through the iterative Scientific Cycle, but not down the road of infinite regress. Intrinsic to this worldview is the idea that we must be perpetually willing to deal with incomplete knowledge, with the challenge being how to deal with such incompleteness in an adequate and sufficient fashion. Such is the case in dealing with the challenges inherent to the issues of patient heterogeneity in the clinical setting; while we are exploring the potential binding mechanisms, we also need to somehow deal with that interindividual variability in the meantime. The solution, strangely enough, comes through capturing randomness or *stochasticity*. To a great degree, stochasticity is an anathema to a good experiment: the design of such experiments is focused on trying to remove as much randomness as possible. Also, stochasticity is often invoked as a proxy for "we do not know why;" i.e., *something* must be causing the seemingly random behavior but we just do not know what it is. At first this seems like a bit of a cop-out, but in actuality it is an important insight that can guide what we can and cannot say about a system. For instance, we can obtain, through experiment, probabilities that one effect or another might occur. We do not know why those probabilities take the form that they do, but they clearly appear to do so in a consistent fashion. Rather than getting stuck trying to find out the source of variability, it is possible to move forward at some level by incorporating this stochasticity into our vehicles of dynamic knowledge representation. This is important at the preclinical level, in order to match experimental observations, but is absolutely critical when performing *in silico* trials. By using stochastic processes within our mechanistic models we can capture the population distributions that are central to evaluating the effect of an intervention, and in so doing offer the possibility of generating useful models abscent complete knowledge. Furthermore, when carried out in conjunction with investigations aimed at identifying those binding mechanistic hypotheses that can potential aspire to becoming theories, the stochastic generation of simulated individuals can provide complementary insights into ranges of responses and behaviors arising from different conditions. This was exactly the case with our initial pair of *in silico* trials of antisepsis therapies [18,19], in which the different approaches (stochastic agent-based modeling versus parameter-exploring ordinary differential equation models) provided complementary insights into why those antimediator trials failed.

Being able to overcome Issue #1 makes addressing Issue #2 relatively straightforward: since we are able to generate simulated patient populations, and dealing with electrons and computers is more scalable and tractable than performing human trials, *in silico* trials can be run at whatever size is necessary to obtain statistical significance (if such a difference is to be found). At one level, this is a bit of a cheat: one can almost certainly run a trial of whatever size is necessary to generate some targeted *p*-value if even some modicum of difference exists. In fact, to some degree, this actually points out some of the potential fallacies associated with relying purely upon statistics. However, alternatively, we suggest that the ability to run nearly unlimited numbers of simulated patients, at levels of resolution not possible in the real world, gives us an unprecedented opportunity to actually try and understand what is going on within the dynamics underlying these population effects. This being the case, the evaluation of an *in silico* clinical trial becomes less an exercise in attempting to achieve statistical or intent-to-treat significance; of course these are desired outcomes of the intervention but, given our recognition of the importance of having a means for "useful failure," it is almost more important that we are able to have a mechanism of being able to assess why a proposed promising intervention did not actually work. This returns us again to the concept of a Randomized Prospective Clinical Trial as an experiment, with part of the experimental design being the planning for how to proceed from a surprising or unanticipated outcome. One example of this approach can be seen in an early modeling project involving the simulation of a series of neutralizing anti-inflammatory mediator antibody interventions, both from existing clinical trials as well as several hypothetical interventions, implemented in an agent-based model (ABM) of systemic inflammation [19,20]. This ABM, which is discussed in greater detail in Chapter 4.4, reproduced the general disease dynamics of sepsis and multiple organ failure, and was used to generate simulated populations corresponding to the control and treatments groups in a sepsis clinical trial. Importantly, these clinical trials were simulated in such a way that assumed that the proposed interventions behaved mechanistically exactly as had been hypothesized. Therefore, these *in silico* trials represented performing conceptual model verification on the rationale for such interventions. In line with actual outcomes, and not surprisingly for those studies that were purely hypothetical, none of the simulated interventions demonstrated a beneficial effect [18,19]. The first conclusion drawn from these findings is that, most likely, the underlying conceptual models that informed the development of these therapeutic strategies targeted at blocking individual mediators were flawed. In addition, being able to closely evaluate the dynamics of each intervention, with specific notice of what actually happened to the system during and immediately after each intervention, provided fundamental insights into the nature of antimediator therapy, specifically that there was a "pebble-in-the-stream" effect where the effect of the intervention was swept away by the dynamic robustness of the underlying system. This study also provided insights into why these patients actually died (hint: it was not because of overly exuberant inflammation!). We will discuss these insights in more detail in Chapter 4.4, but we mention them briefly here as part of our description of the potential benefits of performing *in silico* trials.

Another example of the potential insights obtained from carrying out *in silico* trials can be seen in the early *in silico* trial based on a neutralizing anti-TNF-α (Tumor Necrosis Factor-α) therapy we have mentioned previously [18]. This set of simulations recapitulated the general lack of efficacy of the intervention; however, given the ability to use individual simulated patients as their own controls, it allowed us to evaluate what would have happened in the absence of intervention or in the setting of different doses of the drug. Consequently, this *in silico* analysis suggested specific characteristics of the simulated patients that had been helped by the intervention, had been harmed by the intervention, or had not been affected by the drug, thereby suggesting the possibility of using this *in silico* approach for deciding on inclusion and exclusion criteria for eventual real-world clinical trials. Thus, the key take-home lesson of this study was that a failed randomized, placebo-controlled clinical trial could possibly have been successful through the use of *in silico* modeling.

Finally, our approach to performing *in silico* trials represents a natural extension of our use of dynamic knowledge representation within the Scientific Cycle to test hypotheses more rapidly. To a great degree, our experience in starting from the clinical scenario gives us the opportunity to apply this mode of hypothesis/mechanism testing at the very beginning of the drug development and candidate evaluation pipeline. True, some of the knowledge about how a particular drug might affect distant or normal tissues and organs might not be available at the time of candidate identification, but the ability to contextualize specific pathway effects within a suite of existing, system-level, dynamic models can certainly speed the efficiency and efficacy of their evaluation. Also, this process need not be limited to new drug candidates; as we demonstrated in the early ABM antisepsis simulations, hypothetical interventions as well as potential combination therapies could also be so tested [18,19]. The role of these types of simulations in projects involved in repurposing existing therapeutic agents is an obvious early target for adoption.

Note that, ultimately in their conception, clinical trials are predicated upon the representation of populations of patients, and the use of those populations to determine some statistical standard that defines "efficacy." And so it will likely remain in the future, given the regulatory concerns associated with how the governmental regulatory agencies typically exercise their duties. The ultimate goal is to have the mechanistic models used in the performance of *in silico* trials achieve a level of trustworthiness and engineering rigor seen in other technical industries. In fact, in the area of biomedical device design, which is much more focused on the mechanical–physical properties of these devices while the desired biological interface effect can best characterized as "inert," such modeling and simulation standards appear to be tractable in the relatively near future. This is because the essential gulf between the current state of biological biomedical knowledge and the conditions needed to implement a formal engineering paradigm is that we just don't know enough about biology: the central task here is in *discovery*, rather than *optimization*, which is the goal of engineering. However, by recognizing this gulf, we have taken an engineering approach toward the scientific discovery process in our development of Translational Systems Biology. The development of rational strategies for the performance and interpretation of *in silico* clinical trials is a tangible, useful manifestation of our proposed approach, bringing the power of computation and fundamental scientific soundness of dynamic knowledge representation to bear on a clearly recognizable hurdle to the development of more effective therapeutics. However, part of the effective execution of science involves looking beyond the challenges of today toward the challenges of the future, and Translational Systems Biology accomplishes this goal by looking beyond the current standard of clinical trials toward a future of individually tailored, personalized medicine.

FROM POPULATIONS TO INDIVIDUALS: PERSONALIZING MEDICAL CARE

To paraphrase the warning label on automobile side view mirrors, "The future is closer than it might appear." The theme of "personalized medicine" is one that has been increasingly evident over the past decade, though we will argue that its genesis and current direction fall into the category of a fortuitous false start [21]. More recently, perhaps based on the realization that it may be quite difficult (not to mention expensive) to fully personalize medical care to the individual, the concept of "precision medicine" has emerged [17]. Precision medicine is predicated on the concept that quantitative (typically "omics") data will lead to the definition of patient subgroups, and that these quantitative definitions will replace the existing concept of stage-based diagnosis. We see these strategies as falling back on historically derived behaviors reflecting the descriptive legacy of biology. As we have noted previously, the increasing use of genotyping and Genome-Wide Association Studies (GWAS) to attempt to obtain greater descriptive resolution of individuals is an example of applying the default strategy of biology, in which categorization is the primary endpoint. This approach is a good start, but by now we hope we have made the case that the Scientific Cycle mandates that we need to move beyond this type of correlative pattern matching if we are presuming to attempt to tailor specific therapies to particular individuals. Considerable clinical data might be available regarding the effect of existing therapies within a population, as would potentially be the case for trying to optimize a current drug regimen or in an attempt to repurpose existing drugs. However, the extrapolation of any statistical correlations to a hypothesized effect still requires that any presumed alteration in therapy be tried in the clinical population, which is now, given the goal of personalizing care, collapsed to an experiment with $N = 1$. And this says nothing about the dilemmas involved in trying to develop or implement new therapies: it is the nature of such correlative examinations that there must be an existing data set upon which to operate. Given the challenges we have seen in obtaining those clinical data in the current landscape, we would argue that this is a strategy that is bound to fail.

At this stage in this book, it should be easy to anticipate our solution to the challenge of personalized medicine, particularly since it is just the continuation of the processes of Translational Systems Biology that we have proposed for the preclinical and *in silico* clinical trial phases of the drug development pipeline. In fact, we think that the fact that the overall, general process is conserved through these different phases helps substantiate the fundamental soundness of this approach. Here, we return to the idea of using dynamic mechanistic models as a means of binding together varied observations and behaviors, where the fundamental similarities between biological systems (now in the form of individual patients) are manifest in our dynamic knowledge representations. In fact, the approach by which Translational Systems Biology could solve the issue of clinical patient heterogeneity in the context of *in silico* clinical trials is in essence the exact approach utilized in the development of personalized therapeutic strategies. The key insight we provide is that the ability to create truly personalized simulations is paradoxically reliant upon the ability to generate appropriately generalized and abstract mechanistic models. This returns us to the idealized goal of attempting to find biological theories, but tempered as it must be with the realization that what we are seeking is not an ultimate, universal truth, but rather a sufficient system characterization that can allow for the development and implementation of an effective intervention at the level of the individual.

How is this approach different than attempting to categorize individuals using lists of their particular attributes a particular point in time (i.e., genes and single-nucleotide polymorphisms)? The key point here, as has been our case throughout this book, is the emphasis on representing mechanism and function. The emphasis on characterizing and reifying individuals based on function shifts the frame of description from lists of sequential snapshots of descriptive attributes to model parameters that define differential functions and behaviors. This is an important distinction, because an approach that characterizes biological systems based on parameter values that reflect the particular responsiveness of various biological functions (which is exactly the role parameter values play in mathematical functions) provides a pathway to unifying biological behavior with mechanistic models. While this shifts the nature of the measurements that would be obtained in the evaluation of patients (i.e., a shift from absolute values of mediators or biomarkers to metrics denoting responsiveness, as are currently obtained in oral glucose tolerance tests or steroid stimulation tests) now the observed differences between the behavior and trajectories of individuals would be generated directly from the fundamental mechanistic structure that they share.

We realize that in order to accomplish the goal of using mechanistic models to bind together different biological systems will require several tiers of models, where each tier of abstraction unifies nested groups of biological systems; we certainly do not expect that there will be a set of biological laws akin to those Universal Laws of Motion, Conservation, or Thermodynamics seen in physics and chemistry. In fact, it is our recognition of the distinct possibility that biology is too complex for such a set of fundamental laws that drove us to adopt the idea of "sufficiency" in terms of the "just right" level of detail required for useful predictions from our mechanistic models. Central to this strategy is the realization that the level of representation provided by our mechanistic models is guided by the level of any proposed intervention that model is intended to evaluate: the mechanism to be controlled must be represented at the level of its putative controller. This concept places both upper and lower operational bounds on the level of resolution for our mechanistic models: our models need to be at least finely grained enough such that the effect of the putative intervention does not require multiple assumptions, but at the same time no more detailed than the level of resolution required to represent the targeted mechanism. As an example, we look to a physiological model of shock in the context of critical illness: if the intervention being proposed is the administration of intravenous fluid, then a physiomechanical hemodynamic model of cardiovascular physiology is sufficient to evaluate the hemodynamic responses of fluid administration. Alternatively, a proposed intervention directed at proinflammatory mediators could not use a physiological hemodynamic model. The representation of the mechanistic effects of the intervention on hemodynamics requires a whole series of assumptions concerning the consequences and interactions among complex mechanisms extending from the cellular and molecular effects of the intervention to the physiology of the cardiovascular system. Similarly, a mechanistic model used to evaluate a therapeutic strategy aimed at neutralizing a proinflammatory cytokine does not need to incorporate detailed biomechanical processes associated with alterations in endothelial cell cytoskeletal structure and its altered biophysical properties associated with shock; these effects are operating so far below the presumed intervention that their effective output can be aggregated into an abstracted cellular response to the putative intervention. By utilizing this guided approach to dynamic knowledge representation, we can avoid the trap of infinite regress while having a built-in assessment of whether a posited control strategy could be tested. This is the pathway for developing more mechanistically detailed models within Translational Systems Biology, as opposed to striving for additional detail for detail's sake. Rather, greater mechanistic detail is added to these models only when it becomes apparent that some clinically relevant function or observation is now required, an assessment that usually arises because a more abstract model has been falsified or deemed insufficient. This modeling philosophy reemphasizes the intrinsically iterative approach to the construction of models that mirrors the Scientific Cycle. Our view is that models, like hypotheses, should be considered useful, but disposable objects.

This strategy and approach regarding the path toward personalized medicine are natural outgrowths of the use of dynamic knowledge representation in Translational Systems Biology. More detailed models arise in order to capture increasing distinctions that are made among groups of patients, eventually leading to the treatment of each patient as a unique individual, whose disease manifests as the product of a shared generative mechanism. We assert that there is no other rational path toward the admirable goal of personalized medicine, and further assert that the pursuit of this goal is itself a necessary investigatory direction given the challenges associated with the execution of clinical trials. These two processes of performing *in silico* trials and simulating individual patients, far from mirroring the contradictory position seen between Evidence-Based Medicine and Personalized Medicine, actually represent different aspects of the same solution.

Having provided the rational and overall strategy of Translational Systems Biology, we now turn in Section 4 to the specific tools used to execute this approach to translational research. In this following section, we present multiple examples of how we have already obtained, even at this early stage, significant insights into the Translational Systems Biology of acute inflammation.

References

[1] An G, Nieman G, Vodovotz Y. Computational and systems biology in trauma and sepsis: current state and future perspectives. Int J Burns Trauma 2012;2:1–10.

[2] An G, Namas R, Vodovotz Y. Sepsis: from pattern to mechanism and back. Crit Rev Biomed Eng 2012;40:341–51.

[3] An G, Nieman G, Vodovotz Y. Toward computational identification of multiscale tipping points in multiple organ failure. Ann Biomed Eng 2012;40:2412–24.

[4] Kitano H. Systems biology: a brief overview. Science 2002;295(5560):1662–4.

[5] Csete ME, Doyle JC. Reverse engineering of biological complexity. Science 2002;295(5560):1664–9.

[6] Young SS, Lam RL, Welch WJ. Initial compound selection for sequential screening. Curr Opin Drug Discov Dev 2002;5(3):422–7.

[7] Rovira X, Pin JP, Giraldo J. The asymmetric/symmetric activation of GPCR dimers as a possible mechanistic rationale for multiple signalling pathways. Trends Pharmacol Sci 2010;31(1):15–21.

[8] An G. Introduction of an agent based multi-scale modular architecture for dynamic knowledge representation of acute inflammation. Theor Biol Med Model 2008;5:11.

[9] An G. Closing the scientific loop: bridging correlation and causality in the petaflop age. Sci Transl Med 2010;2:41ps34.

[10] Vodovotz Y, Constantine G, Faeder J, Mi Q, Rubin J, Sarkar J, et al. Translational systems approaches to the biology of inflammation and healing. Immunopharmacol Immunotoxicol 2010;32:181–95.

[11] Vodovotz Y. Translational systems biology of inflammation and healing. Wound Repair Regen 2010;18(1):3–7.

[12] Mi Q, Li NYK, Ziraldo C, Ghuma A, Mikheev M, Squires R, et al. Translational systems biology of inflammation: potential applications to personalized medicine. Per Med 2010;7:549–59.

[13] Vodovotz Y, Csete M, Bartels J, Chang S, An G. Translational systems biology of inflammation. PLoS Comput Biol 2008;4:1–6.

[14] Vodovotz Y, An G. Systems biology and inflammation Yan Q, editor. Systems biology in drug discovery and development: methods and protocols. Totowa, NJ: Springer Science & Business Media; 2009. p. 181–201.

[15] An G, Bartels J, Vodovotz Y. In silico augmentation of the drug development pipeline: examples from the study of acute inflammation. Drug Dev Res 2011;72:1–14.

[16] Vodovotz Y, Clermont G, Chow C, An G. Mathematical models of the acute inflammatory response. Curr Opin Crit Care 2004;10:383–90.

[17] National Research Council (U.S.) Committee on A Framework for Developing a New Taxonomy of Disease Toward precision medicine: building a knowledge network for biomedical research and a new taxonomy of disease. Washington, DC: National Academies Press; 2011. xiii, 128 pp.

[18] Clermont G, Bartels J, Kumar R, Constantine G, Vodovotz Y, Chow C. In silico design of clinical trials: a method coming of age. Crit Care Med 2004;32:2061–70.

[19] An G. In silico experiments of existing and hypothetical cytokine-directed clinical trials using agent based modeling. Crit Care Med 2004;32:2050–60.

[20] An G. Agent-based computer simulation and SIRS: building a bridge between basic science and clinical trials. Shock 2001;16(4):266–73.

[21] Weston AD, Hood L. Systems biology, proteomics, and the future of health care: toward predictive, preventative, and personalized medicine. J Proteome Res 2004;3(2):179–96.

SECTION IV

TOOLS AND IMPLEMENTATION OF TRANSLATIONAL SYSTEMS BIOLOGY: THIS IS HOW WE DO IT

4.1

From Data to Knowledge in Translational Systems Biology: An Overview of Computational Approaches Across the Scientific Cycle

The search for early diagnostics as well as efficacious and safe therapies in general, and for inflammation in particular, has been made extremely difficult by the complexity of the underlying, dynamically integrated processes at the molecular, cellular, organ, and whole-organism levels. Repeatedly, we have introduced the notion that reductionist approaches to such a complex system may be incapable of leading to an understanding sufficient enough to diagnose and cure disease. This problem is made doubly difficult by the massive amounts of data now potentially available to biomedical researchers and physicians who are attempting to study, diagnose, or treat these complex, multifactorial diseases.

We described in Chapter 2.5, the initial attempt to address the issue of "complex systems." This trend percolated into the sepsis/shock research community as well, with several descriptions of how this mindset might be applied to the challenges of research on acute inflammation [1,2]. As was perhaps to be expected, these early editorials followed the viewpoint of the original founders of the Santa Fe Institute, and to a large degree reflected the "Physics Envy" exhibited by the general biomedical research community. As we have noted before, this is not automatically a bad thing, especially if a desire to emulate physics leads to greater rigor, novel methodology, and a paradigm shift toward theory- (and hence model-) driven experimentation. However, as we have also seen in the Complexity Field, this desire to apply some of these less well-defined principles and approaches could lead, paradoxically, to the very opposite of scientific rigor, namely hand-waving, blind faith, and semimysticism. Our challenge in developing a systematic approach to the Translational Dilemma was in being able to leverage the useful and beneficial lessons from the Newtonian Paradigm, while avoiding potential traps by a too close approximation between biology and the physical sciences. Intrinsic to this task is identifying robust and scalable processes that can formalize such beneficial but ineffable aspects of human insight and intuition. Nowhere is this challenge is more important that in the identification of patterns within data.

Let us, for the moment, take a quick detour, and consider the state of data-centric systems and computational biology methods that have emerged as an alternative to reductionist, molecule-, pathway-, and physiologic end-point-centric conceptual frameworks in the field of critical illness. Two, heretofore parallel, approaches have evolved over time in an attempt to address the diagnosis and therapy of acute illness from a systems perspective, both of which utilize multidimensional and multiscale patterns of information. This is the first key point: humans, in some sense, gravitate toward patterns. Our brains are tuned for intuitive recognition of patterns, and there has been some mathematical exploration of whether or not our attempt to formalize scientific concepts runs counter to this innate predisposition [3]. We see patterns everywhere in biomedical research, especially if we look at time courses of multivariate data. We see peaks and valleys in these data streams, and our brains sometimes pick up on—or make up—an association between these time-varying patterns and the outcomes of cells, experimental animals, and patients. It should not be surprising, then, that methods attempting to quantify what we think we see would be developed in order to feed our rational impulses. In fact, an overreliance on our sense that patterns should exist is what Bacon warned about in his description of the Idols of the Tribe: the tendency we have of seeing order and structure because we want to. To a great degree this is the appeal of using rigorous statistical approaches to examining data: there is a presumption of objectivity that goes along with an appeal to the cold formalisms of mathematics. However, as we will see, this is a false perception, as different statistical methods can get you different conclusions, and it is the natural tendency (one that Bacon warned about) for us to find exactly what we are looking for.

So, to turn back to the specific approaches used in the analysis of the critically ill, what we hope for is that the different methods for pattern analysis, while being distinct, can prove to be complementary. As should be evident in our description and discussion of Translational Systems Biology, we emphasize the importance of finding the right tool for a particular task, and that the integration of multiple tools (and the differing perspectives they provide) is what can provide us the greatest effective insight into the processes we hope to control. As such, the different types of data analysis we employ are aimed at elucidating key system properties based on the association of data with other data.

One of these classes of multidimensional data analysis is directly drawn from the realm of physics and engineering. These are *signal processing algorithms*, originally intended to examine the properties of various types of waveforms (e.g., sound, radiation, electronic transmissions) and thus can be used to examine physiologic waveforms (e.g., heart rate variability (HRV)) [4–17]. In particular, there has been a great deal of interest in the multiscale analysis of these waveforms, utilizing advanced metrics that can be used to characterize not only the shape and change of these waveforms, but whether the changes themselves (termed the *variability* of the waveform) fall into any discernable patterns. The resulting output is a series of metrics and numerical indices that quantify the degree of variability in a time-varying data set. Note that this type of analysis is carried out essentially agnostically to what actually generates the waveform signal: this methodology represents a purely phenomenological examination of system-level behavior. While this type of analysis can provide insight into the particular state of the system, it provides no insight into why such patterns should be associated with various conditions.

Why is this distinction important? Let us consider the actual uses of these pattern-oriented analyses of physiological signals. If one wants to diagnose an impending unhealthy situation, one needs, at the very minimum, to analyze the signals obtained from a large number of individuals with (i) the condition under study (let us call it Disease A); (ii) some control population (a difficult enough endeavor, given the high likelihood that individuals of the same age, etc., as the sick patients might suffer from other, related ailments that also would manifest in altered physiologic signals); and (iii) people that are ill with distinct, but still well-defined, ailments of different etiologies from those in group #1 (let us call them Diseases B and C). For each of these groups of individuals, one would need to obtain a large amount of data on physiologic waveforms, and then process these data with multiple algorithms in order to define the metrics that most reproducibly differentiate among these groups of patients. As cumbersome (and expensive) as this purely associative study would be to carry out, it would still not mitigate multiple concerns, including the fact that the characteristic changes in physiological waveforms might be different within the same individual as a function of time (time relative to their disease progression, time relative to administration of any therapeutic drugs, even time of day that the measurements were made). And, after all of the above is done, and assuming that all of the confounding variables have somehow been addressed or mitigated, we still would have… no mechanistic understanding about the disease affecting the original cohort of patients under study; at best, we would have a reasonably good diagnostic that says a patient has Disease A but not Disease B or C (or that the patient is, somehow, not healthy). As one can hopefully infer from the foregoing discussion, in order to move more closely to mechanism we need to turn to mechanism-based techniques of data analysis [18]. We will return to this scenario, and this distinction, again below.

In keeping with the historical emphasis in biology on greater and greater detail in description, this next set of data analysis tools can be classified as pertaining to data mining and bioinformatics. These methods are used in order to analyze molecular patterns (e.g., changes in various molecules measured in tissues or fluids), patterns in cellular images (such as biopsies or microscopic images), and patterns in medical diagnostic images. As we have discussed in our review of the history of biology (Chapters 2.2 and 2.3), this is the creation of different sets of "phenotype panels," consisting of lists of attributes that identify particular biological systems and their various states. We call the resulting output from these methods "data-driven models." The insights provided by these models often take the form of connectivity maps or networks, where associations and correlations between different components (usually reflecting either a direct or inverse relationship in their changes over time) are depicted.

As we have previously noted throughout this book, the Scientific Cycle requires us to turn these correlative patterns into some presumption of mechanism if we are to overcome the Translational Dilemma. Our contention is that mechanistic modeling, with the specified goal of providing dynamic knowledge representation, is the core technology that Translational Systems Biology provides in accelerating the Scientific Cycle, thereby setting into motion a robust and scalable solution to the primary current barriers in biomedical research. As we discussed in Chapter 3.2, this class of models simulated hypothesized biological processes (i.e., knowledge) at some level of abstraction sufficient to recapitulate some or all of the data that were used to build the models, but by representing the generative processes by which data is produced, can also generate predictions about behaviors of the system under study that were not used in building or training the models. This type of modeling can be done at any level from the molecular to the behavior of individual organisms or populations, and can be used to link these different scales [19,20].

Ideally, since all of these modeling approaches represent distinct phases of the Scientific Cycle, an integrated and comprehensive investigative approach should apply this range of methods together in order to better understand how diseases develop, how they progress, and how they might be diagnosed or cured. Surprisingly (or perhaps, given the societal and organizational barriers we noted in Chapter 2.5, not so surprisingly), there have been only rare instances of the use of these modeling approaches in concert to make sense of the tremendous, and growing, mountain of biomedical research and clinical data being generated in the context of critical illness. Here, we will focus on summarizing the current state of the art of these individual computational tools, setting the stage for a more detailed discussion in the following chapters:

1. *Methods for the analysis of physiologic patterns*: As we noted above, this range of technologies is drawn from a substantial body of work originally developed and used to analyze the properties of waveform signals in physics and engineering. While phenomenological in nature, various metrics can be used to characterize the types of waveform patterns associated with different biological states. For instance, an observation that is fairly common in the course of evaluating both healthy and sick individuals involves key aspects of physiology such as heart rate, breathing rate, and changes in blood pressure. Physicians know that these indices of physiological function change when a person becomes sick. In the late 1980s and early 1990s, the notion emerged that insights into impending illness could be obtained from sophisticated analysis of these physiological signals. Progress in this methodology has allowed for the metrics obtained from these signal processing analyses to provide a characterization of clinically relevant output of organ behavior and state.
2. *Methods for the analysis of molecular patterns*: The multidimensional analysis of molecular/genetic data provides a very detailed characterization of the various molecular and gene expression levels associated with experimental perturbations or with health and disease. These data are analyzed and interpreted, looking for patterns, using data-driven (statistically based) computational models (see Chapter 4.2).
3. *Mechanistic modeling (at the molecular, cellular, tissue/organ, whole-organism, and population levels)*: These methods close the iterative Scientific Cycle by representing knowledge, and in so doing allow for abstracting mechanisms, for encapsulating mechanisms, and for providing the critical mechanistic and unifying linking among states within a biological system, and eventually between biological systems (see Chapters 4.3 and 4.4).

Over the past 30 years or so, these three areas of computational analysis have been applied, on and off, to data obtained from cells, experimental animals, and patients in an attempt to diagnose critical illness and, much more rarely, to suggest novel therapeutic modalities. At one level, this is encouraging progress, for it is now becoming clear that the overall enterprise of biomedical research needs integration of data and models—and thus of researchers, clinicians, and modelers—in order to make the next leap from data to clinically actionable knowledge [21]. We could imagine that the different computational modeling approaches represent complementary viewpoints of the same biological or clinical system, each with its distinct benefits, but individually incapable of providing the global view necessary to truly understand a complex biological system sufficiently well to diagnose early changes in health status and to engineer effective therapies. In this chapter, we summarize the progress in each distinct field and highlight the pitfalls inherent if the *status quo* of fragmentation and disconnection among these disparate approaches is maintained. Though our focus on this book is on the fundamental utility of dynamic mechanistic models in order to transform the Scientific Cycle and drive progress in biomedical research, we recognize that advances in linking data-driven analysis to mechanistic models to signal processing methods will play a vital role in driving innovations in acute illness diagnosis and care; after all, this process of integrating computational methods is a recapitulation of the Scientific Cycle that we rely upon. Our goal is not to give a textbook-level treatment to these methods, though we will cite references for those readers seeking further information as well as providing many examples from our own work in subsequent chapters. Our hope is that, together, the chapters in this section will provide insight into the overall conceptual benefits of Translational Systems Biology as a rational approach to biomedical research.

PATTERNS IN PHYSIOLOGY: IS THERE A "THERE" THERE?

The Intensive Care Unit is a hub of activity centered on bedside monitors that give caregivers critical information about various aspects of the patient's physiology. The rapid deterioration of a critically ill patient is driven by highly nonlinear interactions among the original insult (injury or infection), ensuing inflammation, and physiologic deterioration. Thus, changes in physiological parameters of the patient reflect, to some degree, all of these intertwined processes. It is therefore not surprising that one area of active research in critical illness involves attempts to glean insights into a patient's physiological deterioration based on signals retrievable from bedside monitoring devices, with the goal of processing and interpreting these complex physiological signals. The take-home message is that as a patient becomes more sick, physiologic variability usually decreases. Twenty years of research in this area have led to the identification of metrics representing loss of complexity of physiologic variability in heart rate and breathing patterns. Importantly from a translational point of view, these metrics are now being used for the diagnosis of critical illness, though still in a limited capacity [4–17,22,23].

Consistent with our theme of the paradoxical state of modern biomedical science (i.e., our "Two Cities"), we would posit that this advance in the adoption of signal processing analyses has been a double-edged sword with regard to gaining new knowledge of the mechanisms that underlie critical illness. On one hand, the hope is that such metrics will also provide some insights that could be leveraged to segregate patients into subgroups with similar aspects of disease, and thereby hopefully improve diagnosis and care. For example, HRV data have been used to detect a person's cardiac health status, and these metrics can also give insights into the pathophysiological processes that affect heart rate, including respiration, blood pressure, and temperature [8]. However, as we have described above and despite the demonstrated validity and usefulness of these types of biological patterns and physiological signal analyses, these methods remain primarily phenomenological in nature, in essence connecting physiologic patterns with clinical outcome through the use of statistics [24].

Why is this a less-than-ideal situation? In a word, "mechanism;" simply put, metrics of changes in physiology, in isolation, provide no direct link to the underlying mechanism that drives changes in the resultant physiologic signals. Well, why is this important? Let us do a thought experiment as an extension of the Disease A versus B versus C scenario described above. Imagine that one wanted to be able to engineer a new therapy for critical illness based on changes in physiologic variability as implement currently. One would have to obtain physiologic signals over the clinical course of critical illness in people of multiple ages, of both genders, and with the full spectrum of underlying causes and etiologies. Then, to avoid false-positive diagnoses, one would need to do a similar study on other diseases that manifest with reduced HRV (e.g., heart failure or other aging-related diseases) [25]. Then, extensive statistical analyses would have to be used in order to associate each observed change with each observed. This is hardly an efficient, or scalable, approach. Though it might be tempting to simply write off this example as being highly artificial, the reality is that this scenario embodies, in microcosm, the entire thought process behind a purely data-driven approach to disease. One might imagine that extensive effort and cost would be expended on developing such an approach only to find, upon reaching the intended goal, that there is no obvious way by which to translate all of these data into a concept for a therapy. Moreover, this approach is not particularly scalable even if the endgame did not center on the development of new therapies but was focused purely on the development of better disease diagnostics. The Disease A versus B versus C scenario outlined above hopefully makes it clear that the combinatorial complexity of disease and patient phenotypes would render extremely difficult a purely pattern-driven strategy to define a highly sensitive and specific diagnostic, as well.

PATTERNS OF MOLECULES

As is the case for physiology-based data, various sets of biological and clinical data can be considered patterns. As a brief reminder, information in biological systems is coded in DNA, which must be transcribed to messenger RNA (mRNA), which in turn must be translated into proteins, some of which function as enymes that produce various metabolites. Any of the above steps can malfunction in disease, and so changes in the patterns of DNA, mRNA (as well as micro-RNA, a more recently appreciated regulatory form of RNA), proteins (and their various modifications), as well as the ultimate metabolites that are produced can be considered disease biomarkers. At the genomic level, one is observing differences in underlying DNA sequences that could indicate either mutations or changes to DNA that affect how DNA is expressed as mRNA. At the transcriptomic level, one is examining the sum total of expressed mRNA of a given cell or tissue [26–29]. At the proteomic level, the questions concern the sum total of proteins in a given cell or tissue [30–33]. At the metabolomic level, one is concerned with the sum total of all small molecules produced by functional proteins, such as amino acids or sugars in a given cell, biofluid, or tissue [32–34].

The growing number of these "omic" methodologies has resulted in a "data deluge" [35], since these data represent a vast amplification relative to the underlying biolgoical or clinical system. Consider a given biological experiment, in which there might be 3–5 replicates of an experiment that, at most, may consist of five time points of observation. If one examines the level of expression of approximately 30,000 genes (i.e., 30,000 mRNA transcripts) at each time point and from each replicate, the data per time vastly outnumber the underlying observations from which the data are obtained. Moreover, the tools that researchers have to sift through these data often add to the confusion rather than reducing it. Why? The methods of choice for analysis of these "omics" data are invariably based on statistical associations [36–43]. Such analyses may suggest principal characteristics of a given complex, dynamic, multidimensional process, thereby allowing the researcher to visualize the degree of difference among experimental groups, for example, see Refs. [43–46]. While this type of analysis is important it is usually not used to determine what it is about a given data set that makes it similar to, or different from, another data set. Alternatively, "omics" data analysis may define the interconnected networks of mediators and signaling responses that characterize a complex biological process [43]. This type of analysis is usually carried out in parallel with so-called pathway analysis, which is a tool that compares the aforementioned data to a curated database of interactions, derived from the scientific literature [47]. A related area of active research involves data-based or data-driven modeling approaches that do not rely on *a priori* knowledge of the internal state of the system, but rather on input–output data measured directly on the system [48–56] (see Chapter 4.2). However, in order to gain mechanistic insights necessary for the rational design and development of therapeutics, and also potentially for the next generation of diagnostic applications, we suggest that a mechanistic, dynamic characterization of the relevant cellular and molecular mechanisms is required [57–60].

Data-driven modeling tools such as principal component analysis can suggest independent drivers of complex biological phenomena [43–46] (see Chapter 4.2), and the results of these models can serve as the basis from which to derive key modules of mechanistic mathematical models [46] (see Chapter 4.3). Network-based models can suggest how multiple, presumably related, variables interact with each other across individuals, across time, or both [43,61]. If it is possible to obtain data on individuals, these data-driven modeling approaches allow modeling and monitoring dynamic changes (in real time) on an individual basis, in essence comprising a novel class of biomarkers [62]. We have applied this approach in our characterization of systemic inflammation arising from trauma and hemorrhagic shock. As with other more traditional approaches to attempts at developing therapeutic targets, we suggest that there is a discrepancy in world views encompassed in the reductionist versus systems-based approach to disease leads to the design of drugs that are targeted to ostensibly diagnostic symptoms rather than to underlying causes of the disease as a whole. A highly linear (direct chains of cause-effect) view of the biological pathways is presumed to underlie the various discrete symptoms, leading to the generation of drugs without any consideration of impact on other pathways, cells, tissues, and organs. We have the belief, which appears to be substantiated by the recent history of effective/noneffective drug development, that this linear concept of disease is insufficient for characterizing the "systems" level diseases such as systemic inflammation, cancer and autoimmune disordered. Therefore, we have adopted an approach seeking to characterize disease states as more than lists of abberrant mediator levels or physiological disturbances. Toward that end, a recurring theme of studies from our group [61,63,64] and others [65] is that "the network is the biomarker." We suggest that conceptual shift in the focus for intervention, which can now be thought of as "the network is the drug target." In the setting of inflammation, this appraoch was pioneered by several investigators at Harvard/MIT, who obtained high-content data on cells subjected to inflammatory stress in cell culture and showed the presence of central signaling mechanisms [44,66–72]. We extended this work in characterizing the early responses to trauma and hemorrhage, pediatric acute liver failure (PALF), and role of concurrent spinal chord injury on the systemic response to blunt trauma; we will describe this work in more detail in Chapter 4.2.

Despite these advances and insights, there are also important limitations to be taken into account when applying these data-driven modeling approaches. These approaches, by definition, rely on available data and as such are dependent on the quality of the sampled data [73], and cannot deal with physical system interconnections [74]. Hence, these methods do not provide any direct mechanistic information about the system; rather they are based on association among data variables [49,75]. However, despite these limitations, the results of data-driven modeling can provide the initial insights necessary to either generate hypotheses that can be tested experimentally, or to first create mechanistic models from hypotheses can be generated.

The ultimate goal of Translational Systems Biology, however, is to control a given biological system to positively affect human health. This requires a mechanistic understanding, and therefore necessitates the construction of mechanistic computational models. This class of models is predicated on mechanistic models, wherein biological and physiological laws and system interconnections are encoded explicitly, though with varying degrees of abstraction. In Chapters 4.3 and 4.4, we will discuss a comprehensive set of mechanistic modeling studies aimed at gaining novel insights into the pathophysiology of acute inflammation in the setting of critical illness, ranging from the cellular and molecular level all the way to the individual and the population.

As we noted earlier, all these different methods provide complementary components of an integrated investigatory process that recapitulates the fundamentals of the Scientific Cycle. Futhermore, given the nature of the Translational Dilemma as it occurs in an increasingly dense data environment, issues of scalability represent a critical target for any proposed strategy. We have recognized that to a great degree the current Translational Dilemma is heavily generated by a throughput problem at the level of mechanistic hypothesis evaluation (with the Popperian paradigm of falsification particularly releveant in the biomedical arena). While we see (and emphasize) the key role to be played by mechanistic modeling in addressing the current bottleneck, we also recognize that the construction of such models is currently a time-consuming, resource-expensive, and expertise-limited process. As such the implementation of our conceptual solution to the Translational Dilemma has in itself an intrisic operational bottleneck. This realization allows us to propose, relatively early in its development, a solution within the context of Translational Systems Biology: the semiautomation of mechanistic model construction. We describe this approach in Chapter 4.5, as well as the greater implications for the scientific community and the investigatory process as a whole suggested by this solution. In so doing we hope to, ourselves, close the loop in our description of Translational Systems Biology and how it can help address the current challenges in applied biomedical research.

References

[1] Buchman TG, Cobb JP, Lapedes AS, Kepler TB. Complex systems analysis: a tool for shock research. Shock 2001;16(4):248–51.
[2] Tjardes T, Neugebauer E. Sepsis research in the next millennium: concentrate on the software rather than the hardware. Shock 2002;17(1):1–8.
[3] Hong FT. The role of pattern recognition in creative problem solving: a case study in search of new mathematics for biology. Prog Biophys Mol Biol 2013;113(1):181–215.
[4] Godin PJ, Buchman TG. Uncoupling of biological oscillators: a complementary hypothesis concerning the pathogenesis of multiple organ dysfunction syndrome. Crit Care Med 1996;24(7):1107–16.
[5] Annane D, Baudrie V, Blanc AS, Laude D, Raphael JC, Elghozi JL. Short-term variability of blood pressure and heart rate in Guillain–Barre syndrome without respiratory failure. Clin Sci 1999;96(6):613–21.
[6] Korach M, Sharshar T, Jarrin I, Fouillot JP, Raphael JC, Gajdos P, et al. Cardiac variability in critically ill adults: influence of sepsis. Crit Care Med 2001;29(7):1380–5.
[7] Piepoli M, Garrard CS, Kontoyannis DA, Bernardi L. Autonomic control of the heart and peripheral vessels in human septic shock. Intensive Care Med 1995;21(2):112–9.
[8] Fairchild KD, Saucerman JJ, Raynor LL, Sivak JA, Xiao Y, Lake DE, et al. Endotoxin depresses heart rate variability in mice: cytokine and steroid effects. Am J Physiol Regul Integr Comp Physiol 2009;297(4):R1019–R1027.
[9] Heart rate variability. Standards of measurement, physiological interpretation, and clinical use. Task Force of the European Society of Cardiology and the North American Society of Pacing and Electrophysiology. Eur Heart J 1996;17(3):354–81.
[10] Kleiger RE, Stein PK, Bigger Jr JT. Heart rate variability: measurement and clinical utility. Ann Noninvasive Electrocardiol 2005;10(1):88–101.
[11] Pomeranz B, Macaulay RJ, Caudill MA, Kutz I, Adam D, Gordon D, et al. Assessment of autonomic function in humans by heart rate spectral analysis. Am J Physiol 1985;248(1 Pt 2):H151–3.
[12] Barnaby D, Ferrick K, Kaplan DT, Shah S, Bijur P, Gallagher EJ. Heart rate variability in emergency department patients with sepsis. Acad Emerg Med 2002;9(7):661–70.
[13] Chen WL, Kuo CD. Characteristics of heart rate variability can predict impending septic shock in emergency department patients with sepsis. Acad Emerg Med 2007;14(5):392–7.
[14] Pontet J, Contreras P, Curbelo A, Medina J, Noveri S, Bentancourt S, et al. Heart rate variability as early marker of multiple organ dysfunction syndrome in septic patients. J Crit Care 2003;18(3):156–63.
[15] Ahmad S, Ramsay T, Huebsch L, Flanagan S, McDiarmid S, Batkin I, et al. Continuous multi-parameter heart rate variability analysis heralds onset of sepsis in adults. PLoS One 2009;4(8):e6642.
[16] Magder S. Bench-to-bedside review: ventilatory abnormalities in sepsis. Crit Care 2009;13(1):202.
[17] Preas HL, Jubran A, Vandivier RW, Reda D, Godin PJ, Banks SM, et al. Effect of endotoxin on ventilation and breath variability: role of cyclooxygenase pathway. Am J Respir Crit Care Med 2001;164(4):620–6.
[18] Aerts JM, Haddad WM, An G, Vodovotz Y. From data patterns to mechanistic models in acute critical illness. J Crit Care 2014.
[19] Ayton GS, Noid WG, Voth GA. Multiscale modeling of biomolecular systems: in serial and in parallel. Curr Opin Struct Biol 2007;17(2):192–8.
[20] Hunt CA, Ropella GE, Lam T, Gewitz AD. Relational grounding facilitates development of scientifically useful multiscale models. Theor Biol Med Model 2011;8:35.
[21] Food and Drug Administration. Innovation or Stagnation: Challenge and Opportunity on the Critical Path to New Medical Products. 2004 March 2004.
[22] Gang Y, Malik M. Heart rate variability in critical care medicine. Curr Opin Crit Care 2002;8(5):371–5.
[23] Moorman JR, Lake DE, Griffin MP. Heart rate characteristics monitoring for neonatal sepsis. IEEE Trans Biomed Eng 2006;53(1):126–32.

[24] An G. Phenomenological issues related to the measurement, mechanisms and manipulation of complex biological systems. Crit Care Med 2006;34(1):245–6.

[25] Goldberger AL, Amaral LA, Hausdorff JM, Ivanov P, Peng CK, Stanley HE. Fractal dynamics in physiology: alterations with disease and aging. Proc Natl Acad Sci USA 2002;99(Suppl. 1):2466–72.

[26] Chung TP, Laramie JM, Province M, Cobb JP. Functional genomics of critical illness and injury. Crit Care Med 2002;30(Suppl. 1):S51–7.

[27] Cobb JP, O'Keefe GE. Injury research in the genomic era. Lancet 2004;363(9426):2076–83.

[28] Wurfel MM. Microarray-based analysis of ventilator-induced lung injury. Proc Am Thorac Soc 2007;4(1):77–84.

[29] Winkelman C. Inflammation and genomics in the critical care unit. Crit Care Nurs Clin North Am 2008;20(2):213–21. vi.

[30] Nguyen A, Yaffe MB. Proteomics and systems biology approaches to signal transduction in sepsis. Crit Care Med 2003;31(Suppl. 1):S1–S6.

[31] Bauer M, Reinhart K. Molecular diagnostics of sepsis—where are we today? Int J Med Microbiol 2010;300(6):411–3.

[32] Claus RA, Otto GP, Deigner HP, Bauer M. Approaching clinical reality: markers for monitoring systemic inflammation and sepsis. Curr Mol Med 2010;10(2):227–35.

[33] Langley RJ, Tsalik EL, van Velkinburgh JC, Glickman SW, Rice BJ, Wang C, et al. An integrated clinico-metabolomic model improves prediction of death in sepsis. Sci Transl Med 2013;5(195):195ra95.

[34] Serkova NJ, Standiford TJ, Stringer KA. The emerging field of quantitative blood metabolomics for biomarker discovery in critical illnesses. Am J Respir Crit Care Med 2011;184:647–55.

[35] Gough NR, Yaffe MB. Focus issue: conquering the data mountain. Sci Signal 2011;4(160):eg2.

[36] Calvano SE, Xiao W, Richards DR, Felciano RM, Baker HV, Cho RJ, et al. A network-based analysis of systemic inflammation in humans. Nature 2005;437:1032–7.

[37] Liu T, Qian WJ, Gritsenko MA, Xiao W, Moldawer LL, Kaushal A, et al. High dynamic range characterization of the trauma patient plasma proteome. Mol Cell Proteomics 2006;5(10):1899–913.

[38] McDunn JE, Husain KD, Polpitiya AD, Burykin A, Ruan J, Li Q, et al. Plasticity of the systemic inflammatory response to acute infection during critical illness: development of the riboleukogram. PLoS One 2008;3(2):e1564.

[39] Warren HS, Elson CM, Hayden DL, Schoenfeld DA, Cobb JP, Maier RV, et al. A genomic score prognostic of outcome in trauma patients. Mol Med 2009;15(7–8):220–7.

[40] Cobb JP, Moore EE, Hayden DL, Minei JP, Cuschieri J, Yang J, et al. Validation of the riboleukogram to detect ventilator-associated pneumonia after severe injury. Ann Surg 2009;250:531–9.

[41] Qian WJ, Petritis BO, Kaushal A, Finnerty CC, Jeschke MG, Monroe ME, et al. Plasma proteome response to severe burn injury revealed by 18O-labeled "universal" reference-based quantitative proteomics. J Proteome Res 2010;9(9):4779–89.

[42] Zhou B, Xu W, Herndon D, Tompkins R, Davis R, Xiao W, et al. Analysis of factorial time-course microarrays with application to a clinical study of burn injury. Proc Natl Acad Sci USA 2010;107(22):9923–8.

[43] Mi Q, Constantine G, Ziraldo C, Solovyev A, Torres A, Namas R, et al. A dynamic view of trauma/hemorrhage-induced inflammation in mice: principal drivers and networks. PLoS One 2011;6:e19424.

[44] Janes KA, Albeck JG, Gaudet S, Sorger PK, Lauffenburger DA, Yaffe MB. A systems model of signaling identifies a molecular basis set for cytokine-induced apoptosis. Science 2005;310(5754):1646–53.

[45] Namas R, Namas R, Lagoa C, Barclay D, Mi Q, Zamora R, et al. Hemoadsorption reprograms inflammation in experimental gram-negative septic fibrin peritonitis: insights from in vivo and in silico studies. Mol Med 2012;18:1366–74.

[46] Nieman K, Brown D, Sarkar J, Kubiak B, Ziraldo C, Vieau C, et al. A two-compartment mathematical model of endotoxin-induced inflammatory and physiologic alterations in swine. Crit Care Med 2012;40:1052–63.

[47] Khatri P, Sirota M, Butte AJ. Ten years of pathway analysis: current approaches and outstanding challenges. PLoS Comput Biol 2012;8(2):e1002375.

[48] Ljung L. System identification: theory for the user. Englewood Cliffs, NJ: Prentice Hall; 1987.

[49] Janes KA, Yaffe MB. Data-driven modelling of signal-transduction networks. Nat Rev Mol Cell Biol 2006;7(11):820–8.

[50] Young PC. Recursive estimation and time series analysis, 2nd ed. Berlin, Germany: Springer; 2011.

[51] Nicholson AJ. An outline of the dynamics of animal populations. Aust J Zool 1954;2:9–65.

[52] Aerts JM, Berckmans D, Saevels P, Decuypere E, Buyse J. Modelling the static and dynamic response of total heat production of broiler chickens to step changes in air temperature and light intensity. Br Poult Sci 2000;41:651–9.

[53] Ingolia NT, Weissman JS. Systems biology: reverse engineering the cell. Nature 2008;454(7208):1059–62.

[54] Amirpour Haredasht S, Barrios JM, Maes P, Verstraeten WW, Clement J, Ducoffre G, et al. A dynamic data-based model describing nephropathia epidemica in Belgium. Biosyst Eng 2011;109:77–89.

[55] Scheffer M, Bascompte J, Brock WA, Brovkin V, Carpenter SR, Dakos V, et al. Early-warning signals for critical transitions. Nature 2009;461(7260):53–9.

[56] Van LK, Guiza F, Meyfroidt G, Aerts JM, Ramon J, Blockeel H, et al. Prediction of clinical conditions after coronary bypass surgery using dynamic data analysis. J Med Syst 2010;34(3):229–39.

[57] Namas R, Zamora R, Namas R, An G, Doyle J, Dick TE, et al. Sepsis: something old, something new, and a systems view. J Crit Care 2012;27:314e1–314e11.

[58] An G, Namas R, Vodovotz Y. Sepsis: from pattern to mechanism and back. Crit Rev Biomed Eng 2012;40:341–51.

[59] Dick TE, Molkov Y, Nieman G, Hsieh Y, Jacono FJ, Doyle J, et al. Linking inflammation and cardiorespiratory variability in sepsis via computational modeling. Front Physiol 2012;3:222.

[60] Kwan A, Hubank M, Rashid A, Klein N, Peters MJ. Transcriptional instability during evolving sepsis may limit biomarker based risk stratification. PLoS One 2013;8(3):e60501.

[61] Azhar N, Ziraldo C, Barclay D, Rudnick D, Squires R, Vodovotz Y. Analysis of serum inflammatory mediators identifies unique dynamic networks associated with death and spontaneous survival in pediatric acute liver failure. PLoS One 2013;8:e78202.

[62] An G, Nieman G, Vodovotz Y. Toward computational identification of multiscale tipping points in multiple organ failure. Ann Biomed Eng 2012;40:2412–24.

[63] Ziraldo C, Vodovotz Y, Namas RA, Almahmoud K, Tapias V, Mi Q, et al. Central role for MCP-1/CCL2 in injury-induced inflammation revealed by *in vitro, in silico*, and clinical studies. PLoS One 2013;8(12):e79804.

[64] Zaaqoq AM, Namas R, Almahmoud K, Krishnan S, Azhar N, Mi Q, et al. IP-10, a potential driver of neurally controlled IL-10 and morbidity in human blunt trauma. Crit Care Med 2014.

[65] Dehmer M, Mueller LAJ, Emmert-Streib F. Quantitative network measures as biomarkers for classifying prostate cancer disease states: a systems approach to diagnostic biomarkers. PLoS One 2013;8(11):e77602.

[66] Aldridge BB, Haller G, Sorger PK, Lauffenburger DA. Direct Lyapunov exponent analysis enables parametric study of transient signalling governing cell behaviour. Syst Biol(Stevenage) 2006;153(6):425–32.

[67] Aldridge BB, Burke JM, Lauffenburger DA, Sorger PK. Physicochemical modelling of cell signalling pathways. Nat Cell Biol 2006;8(11):1195–203.

[68] Saez-Rodriguez J, Alexopoulos LG, Epperlein J, Samaga R, Lauffenburger DA, Klamt S, et al. Discrete logic modelling as a means to link protein signalling networks with functional analysis of mammalian signal transduction. Mol Syst Biol 2009;5:331.

[69] Cosgrove BD, King BM, Hasan MA, Alexopoulos LG, Farazi PA, Hendriks BS, et al. Synergistic drug-cytokine induction of hepatocellular death as an *in vitro* approach for the study of inflammation-associated idiosyncratic drug hepatotoxicity. Toxicol Appl Pharmacol 2009;237(3):317–30.

[70] Alexopoulos LG, Saez-Rodriguez J, Cosgrove BD, Lauffenburger DA, Sorger PK. Networks inferred from biochemical data reveal profound differences in toll-like receptor and inflammatory signaling between normal and transformed hepatocytes. Mol Cell Proteomics 2010;9(9):1849–65.

[71] Cosgrove BD, Alexopoulos LG, Hang TC, Hendriks BS, Sorger PK, Griffith LG, et al. Cytokine-associated drug toxicity in human hepatocytes is associated with signaling network dysregulation. Mol Biosyst 2010;6(7):1195–206.

[72] Saez-Rodriguez J, Alexopoulos LG, Zhang M, Morris MK, Lauffenburger DA, Sorger PK. Comparing signaling networks between normal and transformed hepatocytes using discrete logical models. Cancer Res 2011;71(16):5400–11.

[73] Wiskwo JP, Prokop A, Baudenbacher F, Cliffel D, Csukas B, Velkovsky M. Engineering challenges of BioNEMS: the integration of microfluids, micro- and nanodevices, models and external control for systems biology. IEE Proc Nanobiotechnol 2006;153:81–101.

[74] Willems JC. The behavioral approach to open interconnected systems. IEEE Control Syst Mag 2007;27:46–99.

[75] Young PC. The data-based mechanistic approach to the modelling, forecasting and control of environmental systems. Annu Rev Control 2006;30:169–82.

CHAPTER

4.2

Data-Driven and Statistical Models: Everything Old Is New Again

The most merciful thing in the world, I think, is the inability of the human mind to correlate all its contents. We live on a placid island of ignorance in the midst of black seas of infinity, and it was not meant that we should voyage far. The sciences, each straining in its own direction, have hitherto harmed us little; but some day the piecing together of dissociated knowledge will open up such terrifying vistas of reality... **H.P. Lovecraft, "The Call of Cthulhu"**

As we have stated elsewhere in this book, systems and computational biology are part of an emerging paradigm for studying complex biological systems in a holistic fashion [1]. The most common data analysis methods in systems biology span a broad range of data-driven and statistical techniques, and these tools have become the mainstay—in some senses, the foundation—of "omics"-related bioinformatics and the field of Big Data. These methods can be categorized roughly into correlative or causative methods, with focus on either learning basic principles of system organization and function [2–4] or building predictive computational models [2,5]. Below, we will discuss key aspects of these methods, and how we have incorporated them into the Translational Systems Biology workflow.

At first glance, and based on our prior discussion, it would appear that data-driven modeling would not be a focus of this book, given the primacy of mechanistic, dynamic knowledge representation in Translational Systems Biology. We have already discussed our concerns that blind overreliance on pure data-driven methods, with the idea of throwing a lot of data at the wall and seeing what will stick, is not the rational approach upon which Translational Systems Biology is predicated. However, our modeling efforts have evolved in concert with our concept of Translational Systems Biology. As part of this evolution, it became evident that in order to carry out our mechanistic modeling more effectively as part of the Scientific Cycle in an increasingly data-rich environment, we had to begin to incorporate data-driven modeling tools in some fashion. However, the main types of data analysis and correlative pattern identification methods being carried out were inadequate or inappropriate for meeting the needs raised by the mechanistic computational models we had developed [6].

As with the different mechanistic modeling approaches (to be covered in Chapters 4.3–4.5), the various types of data-driven modeling methods also have their different capabilities and uses [3]. Unfortunately, as with our experiences with the other aspects of the biomedical research community, there is a nearly complete separation between data-driven and mechanistic modeling, another example of the silos that exist in biomedical research (as we described in more detail in Chapter 2.6). Fortunately, given our experience with working in an interdisciplinary environment that includes both mathematical modelers and biostatisticians [7,8], we realized that we had a unique opportunity to fuse data-driven and mechanistic modeling [9–17] within the explicitly applied goal of Translational Systems Biology. In this chapter, we will discuss the statistical methods that are the underpinning of biomedical research, as well as the mainstay data-driven modeling methods common in systems biology. We will focus on a small subset of these data-driven methods, because we feel they are the best suited to making the connection from highly multivariate, time-dependent data and the dynamic, mechanistic models we seek to generate. We will give examples of insights we obtained into the acute inflammatory response based on these methods, and will finish by examining how data-driven modeling methods can be integrated with mechanistic computational modeling.

TRADITIONAL STATISTICAL APPROACHES TO ANALYZING DATA

Many readers of this book will likely be most familiar with statistically-based approaches. These methods include regression techniques that build models predictive within the conditions of the data on which the models were trained [18]. These statistical models are a mainstay of biomedical research, and an enormous body of literature exists with regard to their use. In and of themselves, these methods cannot provide detailed mechanistic insights. However, they can be used to understand key features of the response under study. These features include nonlinearities (e.g., the response evolves with a trajectory that does not follow a straight line as function of time). Another potentially useful piece of information that statistical methods can provide is the identification of factor interactions that affect the response being studied. In the context of systems biology studies of the acute inflammatory response, we began to use univariate analyses and multivariate regression when we examined interrelationships among circulating inflammatory mediators and the possible predictive value thereof in experimental trauma/hemorrhagic shock in mice [19]. Note that we explored these traditional statistical modeling techniques only after we had already generated multiple mechanistic models of acute inflammation (see Chapter 4.3 for details).

Why so late? Most investigators would have started by obtaining data and carrying out statistical analyses that would demonstrate associations among inflammatory mediators. Given our emphasis on trying to recapitulate clinical settings with a specific goal of mechanistic representation and rational design of potential interventions, we chose to focus our initial modeling efforts on creating a mechanistic framework for integrating well-vetted inflammatory mediators [6]. This starting point is, to a great degree, what makes our concept of Translational Systems Biology so unique, by focusing on the dynamic knowledge representation of existing hypotheses regarding the fundamental nature by which acute inflammation manifests at the clinical level. As such, our mechanistic modeling was an unabashedly biased approach, with the supposition that it was the critical care research community's biases that were driving the then-current direction of drug development.

Alternatively, statistical methods are generally thought of as being unbiased. At some level this gives "data-driven" science a greater patina of objectivity; after all, you do not have to think, but rather "just follow the data." In reality, however, since the different forms of data analysis have their own defined capabilities for, and specific perspectives on, the data being analyzed, the person doing the analysis often chooses the method of analysis and thus may inject bias. Moreover, we wanted to gain mechanistic knowledge of possible translational value. Statistical analyses are usually lacking any insight regarding mechanism, and moreover there is often a temptation to over-fit the models to the data on which they were trained. Why is this a bad thing? An over-fit model (see our discussions regarding Ptolemy, for example) can be quite useful in very select circumstances, and sometimes even for a long period of time. In general, as long as there are no differences in the fundamental dynamics of the system, and no intervention is being planned, data-directed correlative prediction works just fine. However, as we discussed previously, the predictive utility of these types of correlative statistical method can in fact hinder attempts at inferring mechanistic insights, if the standard for what constitutes a "good model" is narrowly defined by its ability to predict. This can be a problem because the mechanistic models may in fact fit the data less well (and in fact they often perform more poorly in this fashion) than the statistical models they are attempting to supplant. As a useful example of this conundrum from our own work, we point to a study in which we utilized both standard statistical models as well as mechanistic, differential equation-based models to study a novel biological pathway in inflammatory cells [20]. In that study, we observed a complex dose- and time course response of these inflammatory cells (macrophages) in cell culture with respect to the induction of this novel pathway. We fit these data quite well with a quadratic equation. This equation—a statistical model—yielded no mechanistic insights whatsoever, but was capable of predicting quite well the results of an experiment at a time point on which the model was not trained (but that occurred within the time frame of the training data set). We also generated a mechanistic, equation-based model that could fit the original experimental data, albeit not as well as the statistical model. We derived novel biological insights from this latter model, but the mechanistic model was not as good as the statistical model at predicting the nonlinear response at the new time point [20]. We could argue that the statistical model was highly over-fit, since it was trained on a relatively large amount of data. Though we could not learn from this model anything more than that the biological response we were studying was nonlinear, nonetheless this model was extremely good at being able to predict the data at the next time point *as long as that next time point occurred within the time frame of the original training data set*.

Statistical models also proved useful in the setting of trauma/hemorrhage in mice. In this study, we were able to show that various mediators could segregate the inflammatory response to hemorrhagic shock from the underlying trauma [19], despite the fact that both this study [19] and our earlier mechanistic modeling work suggested a large overlap between the underlying injury alone versus that same injury combined with hemorrhagic shock [21]. So, the ability to characterize patterns of inflammation biomarkers with a validated statistical model has predictive utility in terms of diagnosis and prognosis, especially if one considers that it is important to detect occult hemorrhage in the setting of trauma [22–24]. Our challenge from this point out is how to close the Scientific Cycle so that we can turn these pattern-oriented insights into actionable interventions. We will present some examples of precisely this process below and in Chapter 4.3.

DATA-DRIVEN MODELING IN SYSTEMS AND COMPUTATIONAL BIOLOGY

Hierarchical Clustering: Sorting What Is Alike from What Is Different

Various associative and machine-learning methods such as hierarchical clustering have been developed relatively recently in order to highlight the natural variability, as well as any overlap, across experimental or clinical conditions, typically in the context of systems biology studies. Indeed, many of these informatics methods form the underpinnings of what is colloquially referred to as "Big Data," to which we have alluded previously. Hierarchical clustering is a relatively simple and unbiased method, which is used in order to see if a given set of data from one individual or class/group closely resemble another individual/group. The degree of similarity is thought of, fairly intuitively, as a "distance": how far apart, in multidimensional space (with a dimension being comprised of one variable being measured) is one class from another. This process is carried out iteratively, generating a branched, treelike classification structure (a dendrogram), which resembles a family tree to some degree. This dendrogram can be used to visualize how similar one class (or cluster) is to another based on closeness of association across multidimensional space. The limitation is the cluster must be built pairwise; since it is purely based on the similarity between the data, the cluster may lack biological relevance [3]. Hierarchical clustering is used extensively in the genomics field.

In the field of critical illness, hierarchical clustering has been used to discern patterns and co-regulated clusters of gene expression associated with sepsis and trauma/hemorrhage in both animals [25–28] and humans [29–33]. The typical hierarchical clustering figure is a heat map, essentially a color map in which low levels are represented in green and high levels are represented in red (with black generally representing zero values). Thus, the net result is very visual: a heat map with a dendrogram on one side showing how the various patterns can segregate defined groups (experimental or clinical, for example).

An ideal hierarchical clustering result would be one in which the data yield a heat map that can be neatly segregated based on experimental grouping or clinical outcome. In our own studies, the opposite is usually observed. Indeed, we utilized hierarchical clustering to highlight the tremendous patient-to-patient variability inherent in inflammatory diseases, for example, in the case of circulating, protein-level inflammation biomarkers in children suffering from pediatric acute liver failure (PALF) [34]. Interestingly, this interindividual variability can be demonstrated using hierarchical clustering of similar inflammation biomarker data in genetically identical mice subjected to experimental hemorrhagic shock [19]. So, you may say that we have used the system against itself: hierarchical clustering demonstrated, very visually, the great heterogeneity inherent in this type of data, and the challenge for sorting through these data using typical statistical methods. We will come back to hierarchical clustering after we discuss another data-driven approach, namely, principal component analysis (PCA).

There have also been various efforts aimed at using statistically based tools that, to a certain degree, reproduce the manner in which a clinician might diagnose a patient's disease state and disease progression dynamically [35,36]. These tools use variants of hierarchical clustering applied to time-varying data, and thus in a sense reproduce the behavior of a "super physician" equipped with access to a tremendous amount of data about the dynamic state of the patient. This is an emerging approach, which may or may not prove useful from a practical diagnostic perspective. In this method, an algorithm examines the baseline conditions of a data set (e.g., the demographics of critically ill patients), and sorts any new patient into one or another cluster based on the prior data available. Then, at a later time point, additional data are obtained (e.g., regarding circulating inflammatory mediators in this new patient). The patient is then clustered again relative to the available data on other patients, but this time based on a combination of the prior demographics as well as the data at this new time point. The process is repeated for as many times as there are time points, tracing out a trajectory in time for the new patient. This is what we mean by simulating a "super physician": a purely data-based way of assessing the change in the condition of the patient as a function of time. This process is hardly mechanistic, but in a sense incorporates mechanism in a vague fashion: presumably, something that is being measured regarding the state of the patient, and that proves useful in describing the current state of the patient, may well be part of the mechanism driving the patient's changing health trajectory. This type of data-driven approach lends itself quite well to being compared against a mechanistic model in which the patient's health trajectory would be simulated as a function of explicit mechanisms; we discuss this latter approach in Chapters 4.3 and 4.4.

Principal Component Analysis (PCA): The First Inklings of Mechanism from Data

Another useful method in the systems/computational biology arena is PCA. This method reduces a high-dimensional data set into a few principal components that account for much of the observed variance in the data. When applied to time-series data, PCA may identify the subsets of the variables under study (genes/proteins/etc.) that are most strongly representative of the response. In our studies of acute inflammation, we have interpreted the inflammatory mediators that comprise these principal components as being the principal drivers of the observed response [3,19]. More recently, in the context of inflammation, we have suggested that the variables inferred using PCA are actually more like intermediates along a continuum of a response. Neither they are the earliest "gears" which determine the trajectory on which a biological process will be set through a combination of positive and negative feedback, nor are they the markers of that ultimate trajectory (which are typical biomarkers), but rather key, "gatekeeper" intermediates. So, methods such as PCA can be quite useful because they help generate mechanistic hypotheses [37]. PCA may also aid in the development of diagnostics by gaining insights into this type of intermediate process, which becomes actionable sooner than the markers of the ultimate trajectory described above. We have previously used PCA to define the circulating inflammatory mediator milieu of mice subjected to hemorrhagic shock [19], the local compartment in which infection is present as well as the blood compartment (in which inflammatory mediators appears due to spillover from the local compartment) in septic rats [37], and the multicompartment interaction of inflammatory mediators in acutely inflamed swine [38]. In Chapter 4.3, we discuss how PCA can be used to sift through a large number of variables in order to determine which of them to use for a mechanistic computational model.

In the near term, PCA may serve as a more efficient or accurate biomarker framework with which to identify the health state of individuals and possibly inform patient-specific interventions [10]. As an unpublished example of this approach, we hypothesized that PCA carried out in a patient-specific fashion could discern early differences in multiple organ dysfunction among otherwise similar trauma patients. To test this hypothesis, we studied two separate subcohorts (approximately 30 patients each) of moderately/severely injured blunt trauma patients. Multiple inflammatory mediators were assessed in serial blood samples from these patients. To define the generalizability of this approach, two different but overlapping assay kits were used for both cohorts. We used PCA to suggest patient-specific, early drivers of systemic inflammation in the form of "inflammation barcodes," followed by hierarchical clustering of PCA-transformed data to define patient subgroups. PCA/hierarchical clustering segregated the patients into two groups (A and B) that differed significantly with regard to both systemic inflammation biomarkers as well as in their degree of organ dysfunction within the first 24 h postinjury and independently of the specific set of inflammatory mediators analyzed. Patients in Groups A and B in both cohorts remained segregated based on MS for up to 5 days postinjury. Multiple inflammatory mediators were significantly different in Groups A and B in both cohorts. Thus, early systemic inflammatory responses posttrauma, though diverse, could be characterized through PCA by defined primary drivers that predispose to different degrees of organ dysfunction.

The take-home message from this study is that while neither PCA nor hierarchical clustering gives us a definite mechanistic understanding, the model created by using these techniques could prove useful in segregating patients whose responses would otherwise not be seen as different from each other. So, as in the Ptolemic case described in Section 1, a model can prove useful even though it is, in the sense of biological mechanism, incomplete (if not wrong!). But why did this method work? A key underlying idea of this study was that though patients are quite different from each other in their *apparent* inflammatory responses, they segregate into "more alike" versus "less alike" responses based, presumably, on their genetic predisposition, specific aspects of their traumatic injury, prior history of inflammatory events, etc. And, in the end, they segregate into only a few outcomes. So, we reasoned that there must be some commonality among some of the patients, and we made the admittedly simplistic assumption that this commonality could be discerned at the level of circulating inflammatory mediators. This highlights our tendency to categorize things, as we discussed in Section 1. It also highlights how a rational, hypothesis-driven approach can be linked to a data-driven modeling method to yield a scientifically interesting and potentially clinically useful outcome. The data alone were insufficient to segregate the patients. We had to make the logical leap that some common inflammatory thread bound some patients more closely to each other than to other patients. We next had to find a computational toolset that would allow us to test this hypothesis.

A key aspect of our approach was to embrace, rather than try to avoid, the issue of patient-to-patient variability in inflammatory responses. Classical statistical tools, and even more modern tools such as hierarchical clustering, are based upon how similar things are to each other. PCA, on the other hand, is predicated on variance in data. It is also particularly suited to data that are collected over time (dynamics, which we covered previously). PCA in essence decomposes a complex time series of data down to a single "barcode" that describes the whole inflammatory process for one patient. We emphasize that principal components, by themselves, often do not lend themselves to direct biological interpretations [3]. Furthermore, a variable (e.g., an inflammatory mediators changing over time) that is naturally very "noisy" may be ranked as important by the PCA method without actually being relevant to the phenomenon under study. Again, this brings into question the very nature of biological data: maybe "noisiness" is a key feature of systems that need to respond continuously to input stimuli. If so, then PCA is indeed a very appropriate technique.

Network Inference Algorithms

Another set of emerging, "quasimechanistic" data-driven modeling tools focus on identifying networks of interaction. Network analysis has been a mainstay of systems biology [39]. These methods are predicted on the notion that biological variables that change in some coordinated fashion over time represent elements of a function program, or, at the very minimum, that these networks are indicative of the dynamic state of the patient. Our focus is on acute inflammation, a compendium of biological events that occur in multiple compartments and at multiple scales of organization from the molecule to the whole organism. We have hypothesized that inflammatory mediators form well-coordinated networks that change over time and that are reflective of the underlying condition of the individual or group of individuals [19,32,40–46]. Networks are defined as consisting of nodes (i.e., variables) that are connected in a directed fashion by edges. These connections can be either simple lines, which just denote an association between the nodes, or arrows, which imply a direction of the association. The compendium of nodes and edges is sometimes also called a "graph." Network properties such as degree of connectivity of any given node, robustness, or total network density/complexity can be quantified, and these metrics can give one a sense of the state of the underlying biological system [47]. Hierarchical clustering and Bayesian methods use high-throughput genomic or proteomic data of several time points and/or conditions to correlate gene expression patterns with function and infer regulatory networks of correlated genes [48,49]. Several developments in these methods over the last 15 years have yielded more informative networks that can be more easily translated into mechanistic models [50,51]. A key point is that any network analysis method must reflect, and yield insights into, the dynamics of a given inflammatory response.

Our first foray into network analysis occurred in the context of attempting to define key interactions in experimental hemorrhagic shock, based on insights from several data-driven methods used in concert [19]. In this study, we learned how a relatively simple networking algorithm can shed insights into the complex, dynamic interplay among inflammatory mediators comes from our first study using network analysis. Most often, network analysis is static: that is to say, extensive data (e.g., the sum total of genes being expressed at the mRNA level) are obtained from multiple subjects or experimental repeats at a single time point, and then the results are represented as a network [32]. This type of analysis, unfortunately, can be misinterpreted in many ways. First, the trajectory of the experimental system or clinical setting is unknown; as a thought experiment, one can draw an infinite number of lines through the single point represented by this network. So, is the network becoming more complex? Less complex? Does it change fundamentally? How similar or different would a network appear, then, when compared across individuals whose disease etiology, especially the starting point, is unclear? To begin to address this point in the context of acute inflammation, we utilized a very simple network discovery algorithm based on correlations (either positive or negative) to discern networks of interactions among circulating inflammatory mediators in mice subjected to trauma/hemorrhagic shock. We carried out this network analysis over multiple time intervals (thus calling the method Dynamic Network Analysis, DyNA) by examining the changes from one time point to the next, and then visualizing highly connected nodes. We also examined nodes representing inflammatory mediators whose levels were elevated but that were not connected to other mediators. Though, as we mention above, data-driven modeling is not in and of itself explicitly mechanistic, we utilized DyNA to examine the difference between injury alone versus the same injury combined with hemorrhagic shock. Based on DyNA, it appeared that an ordered cascade of inflammatory mediator interactions occurred following injury alone. Adding hemorrhagic shock did not merely change the levels of inflammatory mediators. Rather, inflammation in hemorrhaged mice appeared to be elevated yet disconnected for a period of time, which was followed by an apparent attempt to connect multiple mediators. Thus, DyNA suggested a profound qualitative difference between injury alone versus injury combined with hemorrhagic shock [19], this, despite a ~40% overlap in the inflammatory responses in these two experimental paradigms of injury [19,21]. A central feature of the DyNA profile in hemorrhaged mice (but not the animals subjected to injury alone) was the key protein, interleukin-6 [19]. Interleukin-6 plays a multifaceted role in inflammation as well as other aspects of (patho)physiology [52], and this complexity is the subject of ongoing debate with regard to whether this protein is a mediator or a biomarker (or both, or neither) in inflammatory diseases [53–64]. An interesting feature of the DyNA results in our particular study was that interleukin-6 was initially predicted to be a key "elevated but not connected" protein early posthemorrhage, but was later seen to participate in a network of inflammatory mediators [19]. Could this mean that interleukin-6 acts as a biomarker early posthemorrhage, and then transitions to become a mediator, thereby suggesting that the controversy over this issue was merely a matter of not having studied this problem using this type of translationally motivated systems biology approach? This is precisely the type of mechanistic hypothesis that one could generate using a purely data-driven modeling technique, if said technique could be applied in an appropriate fashion to the appropriate question.

More recently, we utilized DyNA to get at yet another part of the Translational Dilemma, namely, the inability to translate cell culture data to clinically meaningful diagnostic or therapeutic targets. We hypothesized that we could use multiple data-driven modeling methods, chief among them hierarchical clustering, PCA, and DyNA, to get at this issue in the specific case of liver inflammation [65]. Liver inflammation is not only important for its own sake; there is longstanding evidence that the liver plays a central coordinating role in the inflammatory response to trauma/hemorrhage, as well as being one of the organs that fail as part of the ensuing multiple organ dysfunction syndrome [66]. It is pretty straightforward to obtain primary liver cells (hepatocytes) from mice and stress them in cell culture by depriving them of oxygen, something that actually happens to the liver in experimental trauma/hemorrhage [67]. We hypothesized that we could subject mouse hepatocytes to hypoxia, assess the production of inflammatory mediators by these stressed cells using data-driven modeling, and thereby identify central nodes/inflammatory mediators. We identified a particular mediator, MCP-1, as apparently central in the inflammatory networks induced in response to hepatocyte stress *in vitro*; the next question was: so what? To see if MCP-1 was also relevant to our original question, namely, in the setting of human blunt trauma, we examined the levels interleukin-6 as a function of MCP-1 using another standard statistical tool (linear regression analysis). Interestingly, it appeared that trauma patients could be segregated based on high versus low circulating levels of MCP-1. When we segregated the patients in this fashion, we found that patients with high levels of MCP-1 had worse clinical outcomes than patients with low levels of MCP-1 postinjury [65]. Without our data-driven modeling of the inflammatory responses of hepatocytes to a relevant—but abstracted—stress, we most certainly would not have picked MCP-1 as a mediator to study from the plethora of mediators we examined. This study again showcases our Translational Systems Biology approach: start with a clinically relevant problem, and attack a limitation in the current biomedical research pipeline in a hypothesis-driven fashion using computational tools as an aid and a guide.

As useful as DyNA was, and continues to be in our studies of inflammation, we continued to seek alternative methods to answer additional questions or to solve other unmet needs. One such unmet need was finding a data-driven method of discerning self-feedback (positive or negative) loops among network nodes. One limitation of DyNA is that this method does not identify such feedback loops, which are a core feature of our mechanistic models of inflammation (see Chapter 4.3). Among network methods, dynamic Bayesian networks (DBNs) are particularly suited for inferring self-feedback based on the probabilistic measure of how well the network can explain observed data, and have been used to gain insights into "omic" data sets [48,50,51,68–72]. DBNs essentially let you see a single network picture that represents an entire time course of data; in this sense, DBNs are similar to PCA. Accordingly, we utilized DBN inference methodology to study the interconnections and possible feedback loops inferred from data on circulating inflammatory mediators in several inflammatory diseases.

In our first study, we examined circulating inflammatory mediators in the setting of PALF [73]. This is a multifactorial, poorly understood pediatric inflammatory disease for which the only current therapy is liver transplantation, and for which there is no vetted scoring or staging system for diagnosis [74,75]. The outcomes in this disease are spontaneous survival (because of self-resolving pathophysiology, and sometimes because a compatible liver is simply not available for transplantation), nonsurvival (death prior to transplantation), and transplantation (which is in essence a man-made outcome, but often conflated with nonsurvival since the assumption is that the patient would have succumbed had a suitable organ not been found). Using hierarchical clustering, we showed that circulating inflammation biomarkers were highly variable among patients, and thus of little utility in segregating among these three outcomes. Using patient-specific PCA followed by hierarchical clustering, we could show that the principal drivers of systemic inflammation in some patients that were spontaneous survivors co-segregated with those that received liver transplants, which suggested the possibility that some transplant patients might have survived even without a transplant. This was already a step forward: data-driven modeling suggested the possibility of segregating some patient subgroups based on their systemic inflammatory responses. Using DBN, we found something much more striking: the networks inferred for spontaneous survivors and liver transplant recipients were nearly identical to each other. In contrast, the network inferred for nonsurvivors were related to, but quite different from, the networks inferred for other two groups. This was an epiphany: networks could be biomarkers; other investigators were reaching the same conclusions based on completely different studies [76]. Moreover, the networks we discerned suggested possible inflammatory control mechanisms that were skewed in particular direction in the nonsurvivors versus the other two groups [34]; this type of quasimechanistic insight is simply not possible with standard, statistically based biomarker analyses.

In our next study, we used DBNs to compare the systemic inflammatory responses of traumatic spinal cord injury (TSCI) patients versus similarly injured blunt trauma patients without TSCI [73]. We utilized stringently matched cohorts that were originally from two separate clinical studies performed at the same medical center. DBN inference pointed to a key difference between these two patient cohorts, namely, a central node consisting of the inflammatory mediator IP-10 [73]. Reasoning that even the non-TSCI patients had some degree of neural damage, we sought to determine if IP-10 could segregate blunt trauma outcome groups. Indeed, it turned out that patients with high circulating IP-10 levels had worse outcomes than patients with low levels of this mediator. As in the DyNA study described above [65], we would not have zeroed in on IP-10 as a potential biomarker of adverse outcomes without having first carried out DBN inference.

The above studies, as well as several other studies in hepatocytes as well as mice subjected to trauma/hemorrhage, were remarkably similar with regard to the DBNs inferred. This finding suggested to us the presence of a conserved network architecture that pointed to the liver as a compartment key to the control of noninfectious systemic inflammation. In these diverse settings across species, cell type, and compartment (blood vs. tissue vs. cells), DBN inference repeatedly pointed to what looked like a two- or three-way switch among inflammatory mediators including MCP-1 and IP-10 (mentioned above) as well as another mediator, MIG. Recall that this hypothesis is derived strictly from data-driven modeling, based on data at the protein level. In Chapter 4.4, we will discuss how we tested this hypothesis by constructing a mechanistic model and then verifying key aspects of the model's prediction using data from blunt trauma patients.

Utilizing similar methods, we attempted to define and contrast dynamic networks of inflammation in blunt trauma without spinal cord injury versus blunt trauma combined with spinal cord injury [73]. Given that spinal cord injury is the largest manifestation of neural disruption in trauma, and that the nervous system is a central regulator of inflammation [77,78] and hence of clinical outcomes such as organ dysfunction, we also sought to determine if key biomarkers inferred from this comparison could segregate the outcomes of trauma patients that did not have spinal cord injury. Using DBN inference, we showed that the dynamic, systemic inflammatory responses of spinal cord injury patients are altered significantly relative to those of blunt trauma patients with otherwise similar injury and demographic characteristics. Furthermore, we implicated the chemokine IP-10/CXCL10 and the cytokine IL-10 in this phenomenon, suggesting that IP-10 and IL-10 production is driven by neural control. We then extended our study to show that IP-10/CXCL10 and IL-10 may serve as biomarkers of adverse outcomes postinjury even in the absence of spinal cord injury [73].

Having employed DBN inference successfully in several clinical studies to suggest principal regulatory mechanisms and to define novel outcome biomarkers, we next started to examine the possibility that we could use DBN inference to gain higher-order insights into the inflammatory response *in vivo*. One hypothesis that we had made essentially at the outset of our Translational Systems Biology work was that inflammation and organ dysfunction form a forward-feedback loop (inflammation→dysfunction→inflammation) [9,16,79]. Since DBN inference can point to self-feedback, we sought to determine if this methodology would point to the presence of this forward-feedback loop *in vivo*. We opted to study an experimental paradigm of sepsis and multiple organ dysfunction in swine, and carried out DBN inference on data that included both inflammatory mediators and parameters of physiological (dys)function [80]. The results were striking: the algorithm did, in fact, infer the presence of such an interaction [80].

STATISTICAL AND DATA-DRIVEN MODELING: A PLACE FOR BIG DATA IN TRANSLATIONAL SYSTEMS BIOLOGY?

There are some in the physics and complex systems community that have sought to escape the "tyranny of data." It indeed may be argued that there are numerous sources of error in data, and who decides which data are the correct ones to obtain? It seems, at times, that investigators in the Big Data field would like to believe that if you measured everything possible at many time points, data-mining algorithms will help you figure out what is wheat and what is chaff (or, that the chaff is actually useful for something).

We would argue that the Scientific Cycle and the Scientific Method require us as scientists to impose ourselves on the analysis methodology. We should use data-driven modeling tools to help drive us toward an understanding of mechanism. As we have pointed out above, data-driven modeling in and of itself may yield useful translatable knowledge, but only if the right methods are applied to the right data (and this is where the scientist must exert his insights). It is not enough to throw data at the wall and use every available statistical and data-driven analysis tool to help the data stick. In the subsequent chapters, we will detail how mechanistic models are generated, how they are used, and why we believe they are the cornerstone of Translational Systems Biology and hence the new Scientific Cycle.

References

[1] Ideker T, Galitski T, Hood L. A new approach to decoding life: systems biology. Annu Rev Genomics Hum Genet 2001;2:343–72.

[2] Mesarovic MD, Sreenath SN, Keene JD. Search for organising principles: understanding in systems biology. Syst Biol(Stevenage) 2004;1(1):19–27.

[3] Janes KA, Yaffe MB. Data-driven modelling of signal-transduction networks. Nat Rev Mol Cell Biol 2006;7(11):820–8.

[4] Kitano H. Systems biology: a brief overview. Science 2002:1662–4.

[5] Arkin AP, Schaffer DV. Network news: innovations in 21st century systems biology. Cell 2011;144(6):844–9.

[6] Vodovotz Y, Clermont G, Chow C, An G. Mathematical models of the acute inflammatory response. Curr Opin Crit Care 2004;10:383–90.

[7] Vodovotz Y, Clermont G, Hunt CA, Lefering R, Bartels J, Seydel R, et al. Evidence-based modeling of critical illness: an initial consensus from the Society for Complexity in Acute Illness. J Crit Care 2007;22:77–84.

[8] An G, Hunt CA, Clermont G, Neugebauer E, Vodovotz Y. Challenges and rewards on the road to translational systems biology in acute illness: four case reports from interdisciplinary teams. J Crit Care 2007;22:169–75.

[9] Vodovotz Y, An G. Systems biology and inflammation Yan Q, editor. Systems biology in drug discovery and development: methods and protocols. Totowa, NJ: Springer Science and Business Media; 2009. p. 181–201.

[10] Mi Q, Li NYK, Ziraldo C, Ghuma A, Mikheev M, Squires R, et al. Translational systems biology of inflammation: potential applications to personalized medicine. Per Med 2010;7:549–59.

[11] An G, Bartels J, Vodovotz Y. In silico augmentation of the drug development pipeline: examples from the study of acute inflammation. Drug Dev Res 2011;72:1–14.

[12] Namas R, Zamora R, Namas R, An G, Doyle J, Dick TE, et al. Sepsis: something old, something new, and a systems view. J Crit Care 2012;27:314e1–314e11.

[13] An G, Nieman G, Vodovotz Y. Computational and systems biology in trauma and sepsis: current state and future perspectives. Int J Burns Trauma 2012;2:1–10.

[14] An G, Namas R, Vodovotz Y. Sepsis: from pattern to mechanism and back. Crit Rev Biomed Eng 2012;40:341–51.

[15] An G, Nieman G, Vodovotz Y. Toward computational identification of multiscale tipping points in multiple organ failure. Ann Biomed Eng 2012;40:2412–24.

[16] Vodovotz Y, Billiar TR. In silico modeling: methods and applications to trauma and sepsis. Crit Care Med 2013;41:2008–14.

[17] Azhar N, Mi Q, Ziraldo C, Buliga M, Constantine G, Vodovotz Y. Vodovotz Y, An G, editors. Integrating data driven and mechanistic models of the inflammatory response in sepsis and trauma. New York, NY: Springer; 2013.

[18] Mac Nally R. Regression and model-building in conservation biology, biogeography and ecology: the distinction between—and reconciliation of—"predictive" and "explanatory" models. Biodivers Conserv 2000;9(5):655–71.

[19] Mi Q, Constantine G, Ziraldo C, Solovyev A, Torres A, Namas R, et al. A dynamic view of trauma/hemorrhage-induced inflammation in mice: principal drivers and networks. PLoS One 2011;6:e19424.

[20] Zamora R, Azhar N, Namas R, Metukuri MR, Clermont T, Gladstone C, et al. Identification of a novel pathway of TGF-beta1 regulation by extracellular NAD+ in mouse macrophages: in vitro and in silico studies. J Biol Chem 2012;287:31003–31014.

[21] Lagoa CE, Bartels J, Baratt A, Tseng G, Clermont G, Fink MP, et al. The role of initial trauma in the host's response to injury and hemorrhage: insights from a comparison of mathematical simulations and hepatic transcriptomic analysis. Shock 2006;26:592–600.

[22] Ruffolo DC. Delayed splenic rupture: understanding the threat. J Trauma Nurs 2002;9(2):34–40.

[23] Fouche Y, Sikorski R, Dutton RP. Changing paradigms in surgical resuscitation. Crit Care Med 2010;38(9 Suppl.):S411–20.

[24] Bramos A, Velmahos GC, Butt UM, Fikry K, Smith RM, Chang Y. Predictors of bleeding from stable pelvic fractures. Arch Surg 2011;146(4):407–11.

[25] Chinnaiyan AM, Huber-Lang M, Kumar-Sinha C, Barrette TR, Shankar-Sinha S, Sarma VJ, et al. Molecular signatures of sepsis: multiorgan gene expression profiles of systemic inflammation. Am J Pathol 2001;159(4):1199–209.

[26] Yu SL, Chen HW, Yang PC, Peck K, Tsai MH, Chen JJ, et al. Differential gene expression in gram-negative and gram-positive sepsis. Am J Respir Crit Care Med 2004;169(10):1135–43.

[27] Brownstein BH, Logvinenko T, Lederer JA, Cobb JP, Hubbard WJ, Chaudry IH, et al. Commonality and differences in leukocyte gene expression patterns among three models of inflammation and injury. Physiol Genomics 2006;24(3):298–309.

[28] Edmonds RD, Vodovotz Y, Lagoa C, Dutta-Moscato J, Ching Y, Fink MP, et al. Transcriptomic response of murine liver to severe injury and hemorrhagic shock: a dual platform microarray analysis. Physiol Genomics 2011;43:1170–83.

[29] Prucha M, Ruryk A, Boriss H, Moller E, Zazula R, Herold I, et al. Expression profiling: toward an application in sepsis diagnostics. Shock 2004;22(1):29–33.

[30] Pachot A, Lepape A, Vey S, Bienvenu J, Mougin B, Monneret G. Systemic transcriptional analysis in survivor and non-survivor septic shock patients: a preliminary study. Immunol Lett 2006;106(1):63–71.

[31] Shanley TP, Cvijanovich N, Lin R, Allen GL, Thomas NJ, Doctor A, et al. Genome-level longitudinal expression of signaling pathways and gene networks in pediatric septic shock. Mol Med 2007;13(9–10):495–508.

[32] Calvano SE, Xiao W, Richards DR, Felciano RM, Baker HV, Cho RJ, et al. A network-based analysis of systemic inflammation in humans. Nature 2005;437:1032–7.

[33] Xiao W, Mindrinos MN, Seok J, Cuschieri J, Cuenca AG, Gao H, et al. A genomic storm in critically injured humans. J Exp Med 2011;208(13):2581–90.

[34] Azhar N, Ziraldo C, Barclay D, Rudnick D, Squires R, Vodovotz Y. Analysis of serum inflammatory mediators identifies unique dynamic networks associated with death and spontaneous survival in pediatric acute liver failure. PLoS One 2013;8:e78202.

[35] Cohen MJ, Grossman AD, Morabito D, Knudson MM, Butte AJ, Manley GT. Identification of complex metabolic states in critically injured patients using bioinformatic cluster analysis. Crit Care 2010;14(1):R10.

[36] Zhou B, Xu W, Herndon D, Tompkins R, Davis R, Xiao W, et al. Analysis of factorial time-course microarrays with application to a clinical study of burn injury. Proc Natl Acad Sci USA 2010;107(22):9923–8.

[37] Namas R, Namas R, Lagoa C, Barclay D, Mi Q, Zamora R, et al. Hemoadsorption reprograms inflammation in experimental gram-negative septic fibrin peritonitis: insights from *in vivo* and *in silico* studies. Mol Med 2012;18:1366–74.

[38] Nieman K, Brown D, Sarkar J, Kubiak B, Ziraldo C, Vieau C, et al. A two-compartment mathematical model of endotoxin-induced inflammatory and physiologic alterations in swine. Crit Care Med 2012;40:1052–63.

[39] Samaga R, Klamt S. Modeling approaches for qualitative and semi-quantitative analysis of cellular signaling networks. Cell Commun Signal 2013;11(1):43.

[40] Kunkel SL, Strieter RM, Chensue SW, Basha M, Standiford T, Ham J, et al. Tumor necrosis factor-alpha, interleukin-8 and chemotactic cytokines. Prog Clin Biol Res 1990;349:433–44.

[41] Elias JA, Freundlich B, Kern JA, Rosenbloom J. Cytokine networks in the regulation of inflammation and fibrosis in the lung. Chest 1990;97(6):1439–45.

[42] Miossec P. An update on the cytokine network in rheumatoid arthritis. Curr Opin Rheumatol 2004;16(3):218–22.

[43] Stavitsky AB. The innate immune response to infection, toxins and trauma evolved into networks of interactive, defensive, reparative, regulatory, injurious and pathogenic pathways. Mol Immunol 2007;44(11):2787–99.

[44] Vasto S, Candore G, Balistreri CR, Caruso M, Colonna-Romano G, Grimaldi MP, et al. Inflammatory networks in ageing, age-related diseases and longevity. Mech Ageing Dev 2007;128(1):83–91.

[45] Foteinou PT, Yang E, Androulakis IP. Networks, biology and systems engineering: a case study in inflammation. Comput Chem Eng 2009;33(12):2028–41.

[46] Alexopoulos LG, Saez-Rodriguez J, Cosgrove BD, Lauffenburger DA, Sorger PK. Networks inferred from biochemical data reveal profound differences in toll-like receptor and inflammatory signaling between normal and transformed hepatocytes. Mol Cell Proteomics 2010;9(9):1849–65.

[47] Somvanshi PR, Venkatesh KV. A conceptual review on systems biology in health and diseases: from biological networks to modern therapeutics. Syst Synth Biol 2014;8(1):99–116.

[48] Husmeier D. Reverse engineering of genetic networks with Bayesian networks. Biochem Soc Trans 2003;31(Pt 6):1516–8.

[49] Jiang D, Tang C, Zhang A. Cluster analysis for gene expression data: a survey. IEEE Trans Knowl Data Eng 2004;16:1370–86.

[50] Shah A, Tenzen T, McMahon AP, Woolf PJ. Using mechanistic Bayesian networks to identify downstream targets of the sonic hedgehog pathway. BMC Bioinformatics 2009;10:433.

[51] Rawool SB, Venkatesh KV. Steady state approach to model gene regulatory networks—simulation of microarray experiments. Biosystems 2007;90(3):636–55.

[52] Mihara M, Hashizume M, Yoshida H, Suzuki M, Shiina M. IL-6/IL-6 receptor system and its role in physiological and pathological conditions. Clin Sci (Lond) 2012;122(4):143–59.

[53] Kopf M, Ramsay A, Brombacher F, Baumann H, Freer G, Galanos C, et al. Pleiotropic defects of IL-6-deficient mice including early hematopoiesis, T and B cell function, and acute phase responses. Ann NY Acad Sci 1995;762:308–18.

[54] Opal SM, DePalo VA. Anti-inflammatory cytokines. Chest 2000;117(4):1162–72.

[55] Yudkin JS, Kumari M, Humphries SE, Mohamed-Ali V. Inflammation, obesity, stress and coronary heart disease: is interleukin-6 the link? Atherosclerosis 2000;148(2):209–14.

[56] Fausto N. Liver regeneration. J Hepatol 2000;32(1 Suppl.):19–31.

[57] Naka T, Nishimoto N, Kishimoto T. The paradigm of IL-6: from basic science to medicine. Arthritis Res 2002;4(Suppl. 3):S233–42.

[58] Diehl S, Rincon M. The two faces of IL-6 on Th1/Th2 differentiation. Mol Immunol 2002;39(9):531–6.

[59] Menger MD, Vollmar B. Surgical trauma: hyperinflammation versus immunosuppression? Langenbecks Arch Surg 2004;389(6):475–84.

[60] Kritchevsky SB, Cesari M, Pahor M. Inflammatory markers and cardiovascular health in older adults. Cardiovasc Res 2005;66(2):265–75.

[61] Mastorakos G, Ilias I. Interleukin-6: a cytokine and/or a major modulator of the response to somatic stress. Ann NY Acad Sci 2006;1088:373–81.

[62] Scheller J, Rose-John S. Interleukin-6 and its receptor: from bench to bedside. Med Microbiol Immunol 2006;195(4):173–83.

[63] Knupfer H, Preiss R. Serum interleukin-6 levels in colorectal cancer patients—a summary of published results. Int J Colorectal Dis 2010;25(2):135–40.

[64] Jawa RS, Anillo S, Huntoon K, Baumann H, Kulaylat M. Interleukin-6 in surgery, trauma, and critical care part II: clinical implications. J Intensive Care Med 2011;26(2):73–87.

[65] Ziraldo C, Vodovotz Y, Namas RA, Almahmoud K, Tapias V, Mi Q, et al. Central role for MCP-1/CCL2 in injury-induced inflammation revealed by *in vitro*, *in silico*, and clinical studies. PLoS One 2013;8(12):e79804.

[66] Peitzman AB, Billiar TR, Harbrecht BG, Kelly E, Udekwu AO, Simmons RL. Hemorrhagic shock. Curr Probl Surg 1995;32(11):925–1002.

[67] McCloskey CA, Kameneva MV, Uryash A, Gallo DJ, Billiar TR. Tissue hypoxia activates JNK in the liver during hemorrhagic shock. Shock 2004;22(4):380–6.

[68] Husmeier D. Sensitivity and specificity of inferring genetic regulatory interactions from microarray experiments with dynamic Bayesian networks. Bioinformatics 2003;19(17):2271–82.

[69] Park Y, Moore C, Bader JS. Dynamic networks from hierarchical Bayesian graph clustering. PLoS One 2010;5(1):e8118.

[70] Grzegorczyk M, Husmeier D. Improvements in the reconstruction of time-varying gene regulatory networks: dynamic programming and regularization by information sharing among genes. Bioinformatics 2011;27(5):693–9.

[71] Wang SQ, Li HX. Bayesian inference based modelling for gene transcriptional dynamics by integrating multiple source of knowledge. BMC Syst Biol 2012;6(Suppl. 1):S3.

[72] Godsey B. Improved inference of gene regulatory networks through integrated Bayesian clustering and dynamic modeling of time-course expression data. PLoS One 2013;8(7):e68358.

[73] Zaaqoq AM, Namas R, Almahmoud K, Krishnan S, Azhar N, Mi Q, et al. IP-10, a potential driver of neurally controlled IL-10 and morbidity in human blunt trauma. Crit Care Med 2014;42:1487–97.

[74] Narkewicz MR, Dell OD, Karpen SJ, Murray KF, Schwarz K, Yazigi N, et al. Pattern of diagnostic evaluation for the causes of pediatric acute liver failure: an opportunity for quality improvement. J Pediatr 2009;155(6):801–6.

[75] Sundaram SS, Alonso EM, Narkewicz MR, Zhang S, Squires RH. Characterization and outcomes of young infants with acute liver failure. J Pediatr 2011;159(5):813–8.

[76] Dehmer M, Mueller LAJ, Emmert-Streib F. Quantitative network Measures as biomarkers for classifying prostate cancer disease states: a systems approach to diagnostic biomarkers. PLoS One 2013;8(11):e77602.

[77] Tracey KJ. The inflammatory reflex. Nature 2002;420(6917):853–9.

[78] Dick TE, Molkov Y, Nieman G, Hsieh Y, Jacono FJ, Doyle J, et al. Linking inflammation and cardiorespiratory variability in sepsis via computational modeling. Front Physiol 2012;3:222.

[79] Vodovotz Y, Csete M, Bartels J, Chang S, An G. Translational systems biology of inflammation. PLoS Comput Biol 2008;4:1–6.

[80] Emr B, Sadowsky D, Azhar N, Gatto L, An G, Nieman G, et al. Removal of inflammatory ascites is associated with dynamic modification of local and systemic inflammation along with prevention of acute lung injury: *in vivo* and *in silico* studies. Shock 2014;41:317–23.

4.3

Mechanistic Modeling of Critical Illness Using Equations

This book is predicated on the power of mechanistic computational modeling as a transformative tool for biomedical research, and especially as a cornerstone of a rational way to carry out translational research. In the context of biomedical research, mechanistic computational models are detailed biological, and possibly also physical, descriptions of a system, based on some combination of prior knowledge, data, and the modeler's intuition. Mechanistic computational models can unveil emergent phenomena not immediately obvious from the interactions that are encoded in the model. A multifaceted and multipurpose set of tools for both analysis and simulation in the context of complex biological systems is now available. These models, based on causative interactions, can be constructed as ordinary differential equations (ODEs), partial differential equations (PDEs), rules-based models (RBMs), and agent-based models (ABMs) among other methods (including hybrid methods), and have the advantage of potentially being predictive outside the range of conditions/time points on which they were calibrated [1–4].

We and others have created mechanistic computational models of acute inflammation in sepsis [5–9], endotoxemia [10–24], and trauma/hemorrhage [10,12,25,26]. In large part, these models (using ODE and PDE [this chapter] as well as ABM [Chapter 4.4]) are based on a prototypical progression of inflammation that we describe below in the context of model structure. Some of these models are purely theoretical (e.g., [5–9,11]), while others are based on data either at the protein [10,12,25,26] or at mRNA [14,15,20–22] level. Similar mechanistic models have focused on related diseases such as necrotizing enterocolitis (NEC) [27–30]. In this chapter, we will detail these studies as a means of describing the modeling tools, methods, and, most importantly, thought processes involved in creating and analyzing these models.

MODELING INFLAMMATION USING ODEs

The Tools of the Trade

The longstanding, primary tool for mechanistic computational modeling is differential equations, typically ODE [31], and this methodology has been used for translational applications in the context of acute inflammation and critical illness [31,32]. What are ODE models? In essence, they are groups of interrelated equations that describe the change over time of various variables. These changes can be roughly thought of as what makes the variable go up minus what makes the variable go down. The model consists of the group of equations, and these equations typically cannot be solved for a solution. Rather, they are simulated numerically on a computer, yielding trajectories over time for each of the variables. The ultimate utility of ODE models depends not only on their structure (i.e., What variables do they include? How are those variables affected by other variables?), but also on the strength of interactions among their various components. This latter aspect is determined by rate parameters and initial conditions. Deriving these rate parameters is a nontrivial process, especially in the context of complex models involving many nonlinear interactions [32,33].

We will discuss these issues in the context of the models of inflammation that we have generated. We have utilized ODE modeling to help define the interrelationships among inflammatory mediators in the blood versus other body compartments both within a given individual (organs/tissue) in sepsis and trauma/hemorrhage [7,8,10–13,19,25,26] as well as related diseases such as NEC (a devastating acute inflammatory disease that affects premature infants [27]). We have also used ODE modeling to delineate novel cross-species immune interactions in the setting of the mammalian host for the bloodfeeding vector in malaria [34]. Though we have focused predominantly on abstracting individual patients and generating simulated populations of such individuals (as we discussed in Chapter 3.3), some of the later ODE models we generated were focused at the level of cells and molecular pathways [18,35].

As we mention in Section 2, human progress can be defined in some sense by the progression in the tools we have used. We also made the case that some of the difficulty in which modern biomedical research finds itself is, perhaps, due to blind faith in the power of molecular biology tools. Well, modeling, too, has tools. A key reason for the near-ubiquity of ODE models is the existence of a robust set of analytic tools, which have been developed and used to elucidate key organizational principles of networks (or subnetworks), the properties that explain the dynamics and robustness/sensitivity of a given complex system, and, perhaps most importantly, the critical points of control in the system [31,32,36]. These tools are useful because they can help define the complex interplay among the parts of the whole biological system. We will describe those studies in greater detail below, since they tell our story in the context of the tools we use.

Tools from dynamical systems theory allow identification of the possible steady state(s) of a biological system as well as the dynamics of the system's time evolution. These tools have been used extensively to explain and/or predict diverse behaviors such as bistability, hysteresis, and oscillations in biological systems [37]. Bistability means that there exist two conserved states—technically, local minima—toward which all states of a system will converge depending on parameter values [31]. One key analysis tool for ODE models involves bifurcation diagrams, which can be used to discern the effects of a particular parameter on the possible steady-state behaviors of a system, including bistability. Bifurcation diagrams can also be used to indicate the transition from, for example, a healthy steady state to a pathological one [6,8,11,38]. The relative importance of model parameters can also be quantified by calculating the change in the model output in response to changes in the parameter values using sensitivity analysis [36,39]. These methods are complementary and can help identify key points that affect the behavior of the system being modeled.

Below, we discuss our work on modeling various aspects of acute inflammation using equation-based models along with these tools. We recapitulate our studies in some detail, with a focus not merely on the tools we used, but more so on the rationale for using these tools. If we are successful, this chapter will serve as a general and practical tutorial on how and why to carry out modeling using ODEs. We note that this methodology has been covered in excellent detail elsewhere [31,32], and so would point the reader to those resources as well.

Initial, Abstracted Models of Acute Inflammation

Our work commenced with purely theoretical, "big picture" approaches to modeling acute inflammation, and the toolset associated with equation-based modeling was crucial to these early studies. The modeling work progressed both in breadth and depth from these early studies, leading to increasingly more comprehensive models that could recapitulate and predict features of inflammation in various experimental animal models of acute inflammation. Aspects of this work are described also in Chapter 3.3 (since they relate to *in silico* clinical trials) and in Chapter 4.2 (since they are interrelated with data-driven modeling work). Separately, we carried out complementary agent-based modeling studies, which we detail in Chapter 4.4.

First, however, we should set the stage with regard to the state of the art in ODE modeling of inflammation and immunity at the time we initiated our work (*ca.* 2000). The history of mathematical modeling of immunity dates back to the 1970s [40]. Early, abstracted ODE models of inflammatory responses to bacterial infection began to appear in the late 1980s [41]. In the early 1990s, ODE models of immune responses and the effects of human immunodeficiency virus (HIV) were also published [42]. These models were abstracted and were not focused on critical illness *per se*. However, they set the stage with regard to the general approach we would take in generating our own mathematical models. However, they differed in key respects, as well. The main difference was that these prior models were not really applicable to critical illness, given the intertwined effects of inflammation on organ (patho)physiology. It was precisely this interrelationship that we sought to address in our ODE models, reasoning that only in this way could we account for, and ultimately predict, the health status of the patient (rather than focusing solely on simulating the immune-inflammatory response).

Our first modeling study on the acute inflammatory response in sepsis [6] was predicated on ODE modeling as well as the analysis tools available for this type of modeling. We had just initiated our somewhat quixotic quest to help decipher, using mechanistic modeling, if and how acute inflammation could play a role in the pathogenesis of sepsis and related phenomena. At this juncture, it is important to emphasize the role of interdisciplinary collaboration [43] (see also Chapter 2.5); without the presence of one doctor that treats sepsis patients, one physicist/mathematician, and one inflammation biologist (and, in some of the latter studies, multiple other individuals including researchers in industry), these studies simply could not have taken place. In this first ODE modeling study, we constructed a three-dimensional (i.e., three-variable) ODE model of acute inflammation in sepsis, consisting of a pathogen and two inflammatory mediators. The model reproduced a healthy outcome and diverse negative outcomes, depending on initial conditions and parameters (which we discerned using parameter sensitivity analyses). We analyzed the various bifurcations that characterized the transition among the different outcomes when key parameters were changed, and suggested various therapeutic strategies for sepsis based on these insights. The key finding of this purely theoretical study was that the clinical condition of sepsis can arise from several distinct physiological states, each of which requires a different treatment approach [6]. This conclusion stands in stark contrast to the monolithic view of sepsis by the regulatory agencies, though it is in agreement with the clinical insights from many prior studies [44,45].

This first study on ODE modeling of sepsis set the stage for all of our subsequent studies. Note that no prospective data were obtained for that study; the model used was sufficiently abstracted that multiple biological elements were lumped into individual broad categories (e.g., "pro-inflammatory mediators"). This model highlights an interesting truism of differential equation-based modeling, which might be stated succinctly as "small models, big insights; large models, small insights." Small models that are well constructed hold the promise of driving large insights about system behavior. Large, more biologically realistic models (which we discuss below) allowed us to make much more quantitative predictions, and their robustness to changes in parameter values can be assessed at least partially using parameter sensitivity analysis. However, they are essentially too large to be analyzed using bifurcation analysis. As we present our experience with both small and large ODE models below, it will become apparent that, based on the use case, each approach can provide useful information.

Our next two studies were related both to this first study [6], in that they consisted of qualitative models that were slightly larger than the first [8,11]. These were still "small model" studies, however, and were focused on two distinct but related goals. In the first case [8], we sought to add an explicit anti-inflammatory variable to our initial ODE model described above [6]. In this study, we generated an abstracted, theoretical model for the acute inflammatory response to infection. This ODE model was used to explore the importance of the dynamics of anti-inflammation in promoting resolution of infection and homeostasis. This model also suggested a clinical correlation between model predictions and potential therapeutic interventions based on modulation of immunity by anti-inflammatory agents [8]. How and why did we go about this? Let us first deal with the "why?" Inflammation, as we have mentioned previously, is a complex system (as loaded as that term is), in which a rapid response to infection or injury requires feed-forward behavior. The advantage of such feed-forward behavior is that it allows for rapid ramp-up of responses to a pathogen or injury. The downside to such a system is that it could spin out of control quite easily if left unchecked; hence, there are multiple points of control for the inflammatory response [46]. One key control mechanism involves negative feedback in the form of anti-inflammatory mediators [47,48], and this key control mechanism was absent in our first ODE model [6]. In this manuscript, we described a core structure for acute inflammation that we have utilized—with additional detail on this core set of interactions—in all of our mechanistic modeling studies [49,50]. That structure, in essence, describes how inflammation is interrelated both to the initiating insult (injury or infection) as well as subsequent, abstracted "damage." In the particular framework described in Ref. [8], bacterial infection leads to the nearly simultaneous activation of both pro- and anti-inflammatory mediators. Pro-inflammatory mediators both act to kill the invading pathogens and, importantly, cause bystander damage to the host. Anti-inflammatory mediators suppress the pro-inflammatory mediators, and in essence act to both mitigate inflammation and allow for healing. Also, the pro-inflammatory mediators can be further stimulated by damaged tissue, in effect setting up a feed-forward loop of inflammation→damage→inflammation. As simple as this structure would appear to be at first glance, it leads to complex, nonlinear behaviors. This is the paradigm that we also utilized for *in silico* clinical trials in a larger model described in more detail below [7], in our larger mouse- [10,12,25,26,51] and pig-specific [19] ODE models, as well as some of our ABMs [52–59] described in Chapter 4.4. Colloquially, this schematic could be referred to as the "no free lunch" paradigm: basically, there is almost no simple way to modulate this structure of inflammation and arrive at only positive outcomes. In this initial paper, we explored what this actually means using this relatively simple model. First, we demonstrated that this ODE model could reproduce core behaviors of acute inflammation in response to infection. This model could reproduce the scenario of an individual fighting off infection successfully and returning to a "healthy" state. If the rate of growth of the pathogen would be sufficiently high, however, the model predicted the presence of persistent infections. A bifurcation analysis of this behavior suggested bistability between health and death due to overwhelming infection (septic death), as well as bistability between health and death as a consequence of doing too good a job of clearing the pathogen (but at the cost of bystander damage) [8]. As expected, the presence of anti-inflammatory mediators was a double-edged sword: bystander damage was reduced relative to what would be predicted in the absence of anti-inflammation; however, pathogen-killing efficacy was reduced, as well [8]. Again, no free lunch.

One interesting result that we observed from these studies was that allowing the anti-inflammatory mediators to vary dynamically always resulted in better outcomes than providing a fixed amount of anti-inflammatory mediators (i.e., simulating the administration of an anti-inflammatory mediator to septic patients, a strategy that failed when tried clinically [47,60,61]; see our additional discussion in the context of clinical trials in Chapters 2.4 and 3.3). We also took our first foray into the realm of *in silico* clinical trials by simulating a specific anti-inflammatory therapy, namely one that involves the neutralization of the key pro-inflammatory mediator tumor necrosis factor-α (TNF-α) with antibodies (anti-TNF-α) [7]. Here we utilized a larger ODE model that included a series of pro- and anti-inflammatory mediators that were modeled explicitly and individually, rather than being lumped together into broad categories. We also included a blood pressure variable, since hypotension is a key pathophysiological change that occurs in septic patients. More important, perhaps, was the inclusion of a "damage" term similar to that described above, but which now comprised of multiple effects driven by pro- and anti-inflammatory mediators. The ODE model was a semiquantitative one, since it was calibrated using some biological rate constants and others that were imputed in order to generate a 1000-patient virtual cohort that, upon "infection," would exhibit a "mortality" rate of 35–40% (in line with sepsis mortality at the time [62]). Using this ODE model, we simulated multiple treatment arms consisting of realistic doses of anti-TNF-α by "cloning" the placebo group into each of the treatment arms. This was a novel concept: we could, in essence, observe what would have happened—for the better or for the worse—to individual patients receiving either placebo or treatment. The results were fascinating: some patients benefitted from anti-TNF-α while others were harmed [7]; the overall result suggested no net benefit, in line with the actual clinical trial outcomes [63]. Another important advance presented in this study was the use of the ODE model to predict potential biomarkers and other characteristics of patients that might benefit from anti-TNF-α, versus those that would likely be harmed [7].

In a later study using a modification of this ODE model, we examined the inflammatory response to anthrax [5,9]. In that study, we modified the properties of the invading bacteria in the parent sepsis model [7] to those specific to *Bacillus anthracis*, and simulated the host response to anthrax infection. Two key characteristics of *B. anthracis* that differentiate them from the gram-negative *Escherichia coli* bacteria we simulated in our prior studies [6–8,11] is that these bacteria are gram-positive (and hence induce a lower-degree inflammatory response), and that *B. anthracis* produces three toxins (lethal toxin, protective factor, and edema factor [64]). We simulated treatment strategies against anthrax in a genetically diverse population including antibiotic treatment initiated at various time points, neutralizing antibodies targeting anthrax protective antigen (a vaccine strategy that has been under development for some time [65–70]), as well as a combination of antibiotics and vaccine. In agreement with studies in mice, our simulations showed that antibiotics only improve survival if administered early in the course of anthrax infection. Vaccination that leads to the formation of antibodies to protective antigen was predicted to be anti-inflammatory and beneficial in averting shock and improving survival. However, antibodies to protective antigen alone were predicted not to be universally protective against anthrax infection. Rather, our simulations suggested that an optimal strategy would require both vaccination and antibiotic administration [9]. These conclusions were derived from a series of *in silico* clinical trials, patterned after those performed in the setting of sepsis as described above. Importantly, this study shows the utility of mechanistic mathematical modeling in settings where the area of study poses considerable danger even in the preclinical setting (since extensive and expensive biocontainment is necessary for the study of anthrax), and where clinical studies of efficacy are essentially impossible to perform, ethically.

We could argue that, along with the contemporaneous agent-based modeling study of *in silico* trials in sepsis detailed in the next chapter [5], the foundation of Translational Systems Biology was established because of the need to simulate anti-inflammatory therapies [7]. A key finding from those studies was that randomized administration of anti-inflammatory mediators was not likely to be successful in sepsis [5,7]. There are various directions that, based on ODE modeling studies, could be pursued in order to define a more optimal administration of anti-inflammatory mediators. Our above-described, a simple ODE model [8] suggested that varying the treatment dynamically as a function of changes in the patient's inflammatory response might be a useful strategy. This type of approach falls in the domain of Control Theory, and, indeed, the analysis of ODE models of biological systems can be approached from a control theory perspective. Achieving robustness and efficiency are core principles of both evolution as well as engineering. Negative feedback (as in the form of anti-inflammatory mediators), is a pervasive biological phenomenon and is also a fundamental component of control strategies [71]. An ODE model is, in essence, a state space representation of a control system. Thus, it is possible to decompose the biological system into a control structure and analyze the role of each component using control theoretic tools that characterize their robustness and identify the key mediators that modulate the performance of such a control system [72]. We view systemic inflammation and multiple organ dysfunction in critical illness as a manifesting a "tipping point" behavior that leads to cascading system failure [73], and this may be due to a failure of various control structures at multiple levels of organization to handle these insults.

At essentially the same time, we generated another abstracted model of inflammation that also contained four variables [11]. The goal of this work was to gain insights into the phenomenon of inflammatory preconditioning. The concept of preconditioning, or the effect of past history on a system's behavior, is a critical factor in understanding and characterizing the inflammatory response as a dynamical system. Inflammatory preconditioning highlights the nonlinear nature of the inflammatory response: stimulation with two or more pro-inflammatory stimuli in succession can lead to responses that are equal to, greater than, or lesser than each stimulus in isolation [74]. Preconditioning is also critical in practical terms. First, preconditioning especially tolerance to bacterial products such as endotoxin (lipopolysaccharide), a key "active ingredient" of gram-negative sepsis, may occur clinically in some sepsis patients [75]. The low-level release of endotoxin may be augmented by antibiotic treatment, which may lead to "pulses" of endotoxin [76]. If timed correctly, a therapeutic preconditioning stimulus can be used to augment natural inflammatory protective responses to a subsequent severe insult, or to blunt an overly exuberant (and hence detrimental) inflammatory response. ODEs are particularly useful for mathematically analyzing inflammatory preconditioning, since it is not possible to obtain explicit formulations for the time evolution of all variables associated with the entire inflammatory response. Interestingly, endotoxin tolerance and preconditioning can be observed at the cellular level as well as *in vivo*, and we were able to show, using a very simple ODE model, that simply accounting for the key signaling induced by endotoxin could produce a spectrum of preconditioning phenomena [18].

Bifurcation analysis was also a key tool in a later study in which we created an ODE model of inflammation in the setting of Necrotizing Enterocolitis (NEC), a disease which affects the gut of premature newborns and which includes a systemic inflammation component that shares many features with sepsis [27]. Further, we examined the potential effects of probiotics (bacteria that are thought to be either benign or beneficial, and which have been tested as a therapy for NEC). Using bifurcation analysis, we made multiple predictions regarding the behavior of the inflammatory response in NEC, both in the absence or presence of probiotics. One key finding was that, under some circumstances or with specific characteristics of the bacteria, otherwise beneficial probiotics could actually become harmful [27]. These dichotomous effects may explain why some clinical studies of probiotic administration in NEC have not shown efficacy (or even suggested harm of this approach due to sepsis) [77,78].

In a subsequent ODE modeling study of NEC, we added further molecular details regarding two key receptors for inflammatory products [29]. These two receptors have been suggested to play pro- and anti-inflammatory effects, respectively. An initial temptation was simply to model them as such, essentially "programming the proof," a key mistake in mechanistic modeling. Rather, we modeled the most conserved effects of these receptors, and allowed the results of the simulations under various initial conditions (reflecting different biological scenarios) to allow net pro- and anti-inflammatory effects to emerge [29].

These are further examples of the "no free lunch" paradigm. In this case, "no free lunch" reflects a key principal of mechanistic modeling, namely that models need to be as objective and balanced as possible in depicting the effects of the model variables on each other. In this particular instance, we were able to achieve this prediction because we structured the ODE model to allow, theoretically, for there to be a pro-inflammatory, harmful effect of probiotics (rather than assuming that they would provide only beneficial effects).

One key strategy we employed in the aforementioned studies of NEC, which we also employed in several of the studies described above, was that of model reduction. A system of ODEs becomes complicated very rapidly as the number of equations increases. It can, therefore, be advantageous to attempt to reduce the number of equations to a manageable number by applying a steady-state assumption. This strategy is most appropriately applied to variables that are transient, and is accomplished by setting their derivatives to zero. Model reduction facilitates carrying analyses such as bifurcation analysis. Using model reduction, we found several important behaviors of this ODE model, as related to preconditioning. A central hypothesis of this study was that preconditioning was predicated on the intertwined actions of pro-inflammatory mediators versus either fast-acting or a slow-acting anti-inflammatory mediators [11]. Indeed, studies with this ODE model reproduced multiple, complex outcomes of preconditioning, such as protecting an individual from death induced by a large dose of endotoxin following prior exposure to a low dose of endotoxin or mounting an exaggerated response to a low dose of endotoxin following prior exposure to a similarly low dose [11]. These behaviors are the *sine qua non* of nonlinearity, akin to saying that one plus one does not equal two, but rather can equal 0 or 20, depending on the circumstances.

The predictions of this ODE model were matched qualitatively against previously published data in experimental animals. As important as this type of verification/validation was, it was not as satisfying as carrying out prospective calibration and validation studies. Thus, we embarked on a combined, iterative program of ODE modeling and experimental studies in order to close the loop. We described those studies below.

Models of Acute Inflammation Based on Data from Animals and Humans: Parameter Estimation and Model Ensembles

Contemporaneously with these abstracted ODE models of the acute inflammatory response, we generated a set of larger ODE models that were calibrated against data in experimental paradigms of acute inflammation and critical illness [10,12,25]. Our initial goal was to test the hypothesis that a single set of mechanisms could account for the inflammatory responses of mice to endotoxin (at multiple doses), surgical cannulation trauma, and surgical cannulation trauma combined with hemorrhagic shock/resuscitation [10]. These are three central paradigms of acute inflammation, which to some degree reproduce some of the features of human endotoxic shock and trauma/hemorrhagic shock [79]. At the time we generated this model, the shock research community viewed these paradigms of inflammation as related but distinct; in contrast, we reasoned that inflammation in these different settings had the same basic "wiring diagram," but with different starting conditions. We therefore generated a 15-variable model that was structured much like the simpler four-variable model described above [8]. The larger size of this model was required in order to be able to calibrate the model against experimental data, and to make quantitative prediction. As we stated above, the smaller model was highly abstracted, and hence each variable represented multiple biological entities. That is not to say that the 15-variable model was extremely detailed, far from it. However, this model was generated in order to strike a balance between manageable size (which, as we discuss below, relates to model calibration/parameter estimation) versus biological reality. This is a core issue in mathematical modeling: it is an enterprise fraught with multiple tradeoffs, and one in which it is vitally important to have a clear idea of the goals to be fulfilled by the model.

As a testament to the longevity of this model (see below), multiple subsequent modeling studies with very different purposes and carried out over a decade afterward [12,25,26,51,80] were well served by relatively minor changes to this core model. For example, we knew from the start that entire elements of the immune response were missing from our ODE model. Likewise, physiology was abstracted to just blood pressure, and we included a very abstract term for whole-organism damage/dysfunction [10]. And yet, this model was able to capture the time evolution of multiple circulating inflammatory mediators in mice that were subjected to three different doses of endotoxin, trauma, or trauma/hemorrhage. Perhaps more impressive was the ability of the model to predict a threshold dose for endotoxin-induced death in animals, despite having been trained only on data from animals that did not die [10]. The core finding of this study, though, was that it was possible to reproduce all of these different biological outcomes, over time, in animals, using a single model.

A key reason likely underlying the success of this ODE model at being able to reproduce so many aspects of acute inflammation *in vivo* was the model calibration process. This process is also known as parameter estimation, and basically involves modifying the rate parameters of models so as to better fit a training data set [31,32,49]. There are multiple ways to approach parameter estimation. In an ideal case in a biological setting, the modeler would have access to the exact biological rates, number of cells/molecules, etc. In reality, and especially in experimental animals or patients, this is rarely possible. Knowing this limitation, a model can be structured so as to be able to rely on apparent rates, i.e., the rates that might be actually measurable and which themselves may represent an amalgam of underlying rates and processes. In the case of our initial ODE model of inflammation, we structured the model so as to be calibrated against levels of inflammatory mediators in the blood, which is an easily accessible compartment of the body. Despite the fact that inflammatory processes are typically local, they may still spill over to the systemic circulation when the pro-inflammatory stimulus is sufficiently potent. This "tip of the iceberg" situation clearly can underestimate the magnitude of local inflammation, but as we state above, it is better to have some data than no data. To account for this attenuation, we included special rate parameters in our ODE model [10]. As an aside, we did assess local versus systemic inflammation in septic rats, in the context of a data-driven model, and confirmed this "tip of the iceberg" effect [81].

As a general rule, model parameters are optimized to allow the model *as a whole* to fit the data *as a whole*. Thus, it is possible that for any given scenario, any given time point, or any given variable to fit the data less well than a purely statistical model. Remember that statistical models have little or no mechanistic relevance *per se*, while the types of mechanistic models we are discussing are inherently mechanistic in their structure. Thus, a good fit of a mechanistic ODE model (see below for what that means) to the data raises one's confidence that the model reproduces the referent system (or key features thereof) well.

A good fit to the data is defined by an error function, a numerical index that is at its lowest value when the values of a given set of parameters lead to output of a given set of variables that matches the data as well as possible. "As well as possible" refers to the fit to the data of all other tested values of this given set of parameters. There are multiple methods for carrying out parameter estimation, from relatively simple ones that are often only good at obtaining local minima (e.g., least squares) to very sophisticated genetic algorithms, simulated annealing, and so-called "global" methods that explore the parameter space of a model extensively if not exhaustively. These methods are all algorithms that take initial (seed) values of parameters and change them through random guesses or directed processes; the resulting parameter values are plugged into the model and the model is simulated. The simulation output is then compared against available data, and the discrepancy between the model prediction and the actual data is scored with the error function.

In the case of our initial mouse-specific ODE model [10], the fit was overall quite good between model output and the time courses of four inflammatory mediators that we used for the calibration of this model. We note that we were practical in our selection of data for this process, because we were limited both in funds and by technology available at the time with regard to how many inflammatory mediators we could assay. Here lay both a challenge and an opportunity. The challenge was to find four mediators that would (i) be relevant to the experimental paradigms of acute inflammation we studied, (ii) reflect both pro- and anti-inflammatory processes, and (iii) evolve with distinct time courses. The "Big Data" solution would have been to measure everything possible, and indeed we were asked for quite some years after performing these studies if this is what we did. The complicated solution involved many discussions and choices, and ultimately yielded a model that, as we mention above, has remained useful far beyond the time we would have imagined.

Next, we used this same model to ask a basic question regarding the response to injury and hemorrhagic shock: just how much does the underlying injury contribute to the inflammatory response [25]? We asked this question in the context of the same experimental paradigm of trauma/hemorrhagic shock and ODE model thereof that we employed in our initial study [10]. In order to create experimental hemorrhagic shock, mice must first be cannulated surgically ("sham") so that blood can be withdrawn ("shock"). From a simplistic standpoint, you cannot get blood loss without first damaging tissues and blood vessels, and yet the focus has generally been on hemorrhage rather than the underlying trauma with regard to subsequent inflammation and morbidity [82–84]. Anecdotally, experimenters have appreciated the presence of a so-called "sham effect," which refers to a large overlap in inflammatory mediators when comparing sham- versus shock-treated animals. Our mathematical model predicted that most of the contribution to the acute inflammatory response induced by sham cannulation trauma combined with hemorrhagic shock would be due mostly to the sham cannulation. In fact, we performed an *in silico* experiment that simply cannot be performed *in vivo*: we simulated hemorrhagic shock in the absence of cannulation trauma. This *in silico* experiment suggested a negligible additional effect of hemorrhage on trauma-induced inflammation early postinjury. Experimental studies supported this prediction: circulating inflammatory mediators were essentially identical at the predicted early time point, and the liver transcriptomic response (recall our discussion about the role of the liver in hemorrhagic shock in Chapter 4.2) showed a large overlap between sham and shock animals. Thus, we showed in this study the utility for mathematical models to predict both qualitative and quantitative features of acute inflammation of relevance to critical illness. In addition, this study highlighted another important manifestation of inflammatory preconditioning, following our previous modeling studies in sepsis [11]. As in some of our previous studies, this work also another goal. In this case, we sought to determine if different systems biology approaches, namely ODE modeling and "omic" analysis, would yield similar insights regarding the relative contributions of trauma and hemorrhage to inflammation and tissue dysfunction. Importantly, assessing this transcriptomic required some of the data-driven tools of bioinformatics. Thus, as we describe in somewhat more detail in Chapter 4.2, it is possible to cross-check, as well as integrate, data-driven and mechanistic modeling.

In subsequent studies [12,13,51], we pushed the limits of parameter estimation further by generating populations (ensembles) of ODE models that had the same structure but different sets of parameter values. At this point, it is worth reiterating the old saw that "all models are wrong, but some models are useful." The first part of that oft-used statement refers to the basic truism that all models are abstractions of their referent systems. A corollary to this statement might be "all parameters are wrong, but some parameter sets are useful." As discussed above, parameters in biological models are typically amalgams of multiple underlying processes. The typical ODE model is deterministic, as we mention above. Thus, any single "model" most likely represents only one possible combination of parameter values that could fit the referent data. So, one alternative to deal with this conundrum is to use stochastic ODE models, but as we discuss above this adds a fair bit of complexity to the modeling enterprise. An alternative approach is to generate a series of models that fit the data equally well; this is an approach that is used in applications such as weather forecasting, climate change, and atmospheric particle dispersion [85–90].

In our first ensemble modeling study [12], we sought to understand the *in vivo* effects on acute inflammation of knocking out a gene in mice. As we discuss in Chapter 2.3, the Molecular Biology Revolution has been driven in large part by the ease of generating genetically modified organisms. Of these, gene knockout mice have been a mainstay in the study of acute inflammation and critical illness. However, there are crucial gaps between genotype and phenotype [91], especially in this field [92]. This study also attempted to determine if there was any role for a specific mechanism of endotoxin recognition in the response to trauma/hemorrhage, a matter of some controversy in the field at the time. To address these issues, we hypothesized that we could modify the parameter values of our first mouse-specific ODE model [10] to account for the data obtained in mice in which the gene for a particular endotoxin co-receptor had been knocked out. The results were fascinating: we were able to generate an ensemble of ODE models that fit the data in these gene knockout mice. We used automated algorithms to query the compendium of changes to these parameters that allowed for this very good fit to the resulting data, and this analysis highlighted changes in biologically plausible parameters (given the known functions of endotoxin recognition). Importantly, this analysis also yielded some nonintuitive interactions. With regard to the hypothesis under study, we obtained a good fit to the data in the gene knockout mice without invoking any role for endotoxin in trauma/ hemorrhage, and this was validated with the finding of similar levels of inflammatory mediators in both the wild type and gene knockout mice [12]. In summary, this study showed, using ODE modeling, the potential to address controversies in biology [36]. This study also showed how to integrate molecular biology with modeling; we will discuss this further below in the context of another study in which we interfaced "omics" data with ODE modeling [25]. Perhaps equally important was the establishment of a methodology by which to integrate ensemble modeling into the discovery process: start with a baseline model, obtain data in a new setting or under a perturbation (e.g., gene knockout), recalibrate an ensemble of models to these new data, and query the resulting parameter sets (which represent changes in the underlying biological processes) in order to generate hypotheses about what the perturbation actually does *in vivo*. One could easily imagine this work flow as a means by which to discern what a drug actually does (as opposed to what we think it does, or what it is supposed to do) [93].

A key example of this approach was our use of this work flow to predict key aging-related changes in the acute inflammatory response in mice [51]. In elderly individuals, the inflammatory response becomes radically altered, and is often accompanied by low grade, chronic, inflammation even in the absence of external stimulus [94–96]. Indeed, the linkage between inflammation and aging is close, prompting some authors to coin the term "inflammaging" for this complex process [96]. We hypothesized that we could obtain insights into cellular-level changes characteristic of the aged inflammatory response by examining the systemic responses of "middle-aged" mice, reparameterizing our mathematical model for the levels of circulating inflammatory analytes in these mice, and deriving insights into the cellular and intracellular changes that could lead to these systemic alterations. In this study, we again generated an ensemble of ODE models that were fit to the data obtained in "middle-aged" mice. Our automated analysis of the changes in parameter values that were required for fitting this ensemble to the data suggested key effects on the viability and inflammatory mediator production of macrophages (a central cell type in acute inflammation). We verified these predictions by studying macrophages from young and middle-aged mice. This study again highlights the power of mechanistic modeling—carried out with an explicitly translational goal—to get at central questions affecting health.

Our next study in mice reflected, to some degree, the "Physics Envy" that we have mentioned previously in the biomedical research community [26]. How so? A key aspect of experimental physics is the construction of apparatus in order to test theories (which, in turn, are typically formulated as equation-based models). We had been concerned that our mathematical model of acute inflammation in mice, which could reproduce the inflammatory response to hemorrhagic in mice [10,12,25], was held back by relatively imprecise experimental data. In all of these previous studies, as well as many others published by multiple groups, hemorrhagic shock was performed manually by a human being, with the goal being to reach a target blood pressure through the withdrawal of blood. If too much blood is withdrawn, the person performing the study has to reinfuse shed blood (usually with additional saline resuscitation fluid) in order to reestablish the target blood pressure. This process is imprecise at best, and is quite difficult to model accurately using equations. So, we reasoned that rather than trying to simulate the imprecise actions of a human experimenter in order to match our ODE model predictions to the experimental data, we would build a device that would bleed the mice in the manner that was embodied in the ODE model. So, with the help of collaborators, we built the first-even (and likely only) closed-loop, computer-controlled device for automated hemorrhage and resuscitation in mice [26]. The net result was the ability to model—carry out—very precise hemorrhagic shock. Interestingly, while this approach generally resulted in better fit of model output to the data (including being able to simulate the responses of individual mice), we still observed some variability in the inflammatory responses of individual, genetically identical mice [26]. One hypothesis by which to account for this result comes directly from the structure of the ODE model, namely that individual mice experience slightly different initial conditions of injury, which ripple through the series of positive and negative feedbacks to yield an overall response that is either above a threshold or below it.

What is the implication of this finding, if true? For one, it suggests that the standard, statistics-based analysis of the "average" result for an experimental (or clinical) cohort is not very meaningful. Rather, the individual's response is what matters. To begin to test this hypothesis, we carried out a modeling study in genetically outbred swine challenged with endotoxin [19]. In this study, we also addressed the relative simplicity of our prior, mouse-specific ODE models, in which the entire animal was abstracted as a single, big compartment [10,12,25,26]. Since this particular animal model of endotoxin-induced inflammation results in acute lung injury as well as systemic inflammation, we increased the sophistication of the ODE model by creating two compartments ("lung" and "blood"). We then generated individual-specific ensembles of ODE models using the same methods we describe above [12,51]. The experimental data posed a challenge: animals receiving the same dose of endotoxin spanned the gamut of outcomes from life to extreme sickness to death [19], supporting the hypothesis derived from our mouse study above [26]. Another important aspect of this study in swine was that it represents a fusion of data-driven and mechanistic modeling, in that we utilized Principal Component Analysis (see Chapter 4.2) to define a key module of in the mechanistic model.

Our most recent, unpublished work was focused on individual-specific modeling in the context of simulating the complex, variable responses of individuals and cohorts of blunt trauma patients, based on an extension of this pig-specific ODE model. To address this complexity, we expanded the multicompartment, pig-specific ODE model [19], to include "tissue" (in which physical injury could take place), "lungs" (which can experience dysfunction), and "blood" (representing the circulation as well as a surrogate for the rest of the body), along with inflammatory cells and mechanisms that drive whole-organism "damage." Individual-specific variants of this model were generated from clinical data on 33 blunt trauma survivors. A cohort of 10,000 virtual trauma patients was generated from the 33 patients' individual inflammatory and physiological trajectories, and subjected *in silico* to injury of varying severity. The resulting distributions of model variables equated with length of stay in the intensive care unit, degree of multiple organ dysfunction, and the dynamics of a key inflammatory mediator (interleukin (IL)-6) area under the curve were in concordance with those observed in a separate validation cohort of 147 blunt trauma patients. In the virtual patients, two inflammatory mediators (IL-1β and IL-6) were the main drivers of outcome following mild or moderate/severe injury, and a 50% *in silico* elevation of IL-6 was predicted to convert some survivors to nonsurvivors. In a subcohort of ~100 virtual patients, those with high IL-6 single nucleotide polymorphisms (SNPs) exhibited higher plasma IL-6 levels than those with low IL-6 SNP. Nonintuitively, simulated outcomes in the overall cohort were independent of propensity to produce IL-6, a finding verified in the ~100-patient subcohort. *In silico* randomized clinical trials suggested a small survival benefit of IL-6 inhibition, little benefit of IL-1β inhibition, and, in accordance with our prior studies in trauma patients [97], worse survival following TNF-α inhibition. This study highlights the nonlinear nature of acute inflammation in trauma, and may offer a computational platform for the development of novel diagnostics and therapeutics, both key goals of Translational Systems Biology.

Equation-Based Modeling of Inflammation: What Is in Store?

The foregoing represents a decade-long effort at modeling acute inflammation and critical illness in using ODE models. This work comprises not only a series of scientific studies, but also a journey of self-discovery and insight into how inflammation is wired. We were fortunate to have had the opportunity to collaborate with a large number of co-investigators, many of whom specialize in equation-based modeling. These collaborators were generous with their time, and helped us understand the fundamentals of modeling biological systems with equations. In this chapter, we have tried to blend these insights with the actual methods we utilized, in an attempt to give the reader a sense of how it is that one can incorporate this type of modeling into biomedical research. We are also grateful to other investigators who have taken up the mantle of modeling inflammation using ODE models; their studies have helped further the insights into such aspects of inflammation as the effects of circadian rhythms and the ability to extract key features of inflammation from "omic" data [14,15,21–23,98].

Equation-based modeling is a time-tested methodology for studying complex systems. Elsewhere in this book, we discuss alternative means of mechanistic modeling, but it is very likely that ODE-based modeling will remain a mainstay for many years to come. This is not to say that ODE modeling will remain unchanged or unaffected by these alternative modeling methods. For example, hybrid modeling (blending differential equations with ABMs) is an emerging methodology, which we used in our studies of inflammation and wound healing [58]. We have also mentioned previously the utility of stochastic ODE models, and we have begun to explore this methodology in the context of modeling inflammation. In short, the rich history and tools available for ODE modeling, as well as its extensibility and ability to include multiple compartments, ensure that ODE modeling will remain useful long into the future.

References

[1] Vodovotz Y, An G. Systems biology and inflammation Yan Q, editor. Systems biology in drug discovery and development: methods and protocols. Totowa, NJ: Springer Science and Business Media; 2009. p. 181–201.

[2] Vodovotz Y, Billiar TR. *In silico* modeling: methods and applications to trauma and sepsis. Crit Care Med 2013;41:2008–14.

[3] An G, Faeder J, Vodovotz Y. Translational systems biology: introduction of an engineering approach to the pathophysiology of the burn patient. J Burn Care Res 2008;29:277–85.

[4] Vodovotz Y, Csete M, Bartels J, Chang S, An G. Translational systems biology of inflammation. PLoS Comput Biol 2008;4:1–6.

[5] An G. *In silico* experiments of existing and hypothetical cytokine-directed clinical trials using agent based modeling. Crit Care Med 2004;32:2050–60.

[6] Kumar R, Clermont G, Vodovotz Y, Chow CC. The dynamics of acute inflammation. J Theor Biol 2004;230:145–55.

[7] Clermont G, Bartels J, Kumar R, Constantine G, Vodovotz Y, Chow C. *In silico* design of clinical trials: a method coming of age. Crit Care Med 2004;32:2061–70.

[8] Reynolds A, Rubin J, Clermont G, Day J, Vodovotz Y, Ermentrout GB. A reduced mathematical model of the acute inflammatory response: I. Derivation of model and analysis of anti-inflammation. J Theor Biol 2006;242:220–36.

[9] Kumar R, Chow CC, Bartels J, Clermont G, Vodovotz Y. A mathematical simulation of the inflammatory response to anthrax infection. Shock 2008;29:104–11.

[10] Chow CC, Clermont G, Kumar R, Lagoa C, Tawadrous Z, Gallo D, et al. The acute inflammatory response in diverse shock states. Shock 2005;24:74–84.

[11] Day J, Rubin J, Vodovotz Y, Chow CC, Reynolds A, Clermont G. A reduced mathematical model of the acute inflammatory response: II. Capturing scenarios of repeated endotoxin administration. J Theor Biol 2006;242:237–56.

[12] Prince JM, Levy RM, Bartels J, Baratt A, Kane III JM, Lagoa C, et al. *In silico* and *in vivo* approach to elucidate the inflammatory complexity of CD14-deficient mice. Mol Med 2006;12:88–96.

[13] Daun S, Rubin J, Vodovotz Y, Roy A, Parker R, Clermont G. An ensemble of models of the acute inflammatory response to bacterial lipopolysaccharide in rats: results from parameter space reduction. J Theor Biol 2008;253:843–53.

[14] Foteinou PT, Calvano SE, Lowry SF, Androulakis IP. Modeling endotoxin-induced systemic inflammation using an indirect response approach. Math Biosci 2009;217:27–42.

[15] Foteinou PT, Calvano SE, Lowry SF, Androulakis IP. *In silico* simulation of corticosteroids effect on an NFkB-dependent physicochemical model of systemic inflammation. PLoS One 2009;4(3):e4706.

[16] An G, Faeder JR. Detailed qualitative dynamic knowledge representation using a BioNetGen model of TLR-4 signaling and preconditioning. Math Biosci 2009;217:53–63.

[17] An G. A model of TLR4 signaling and tolerance using a qualitative, particle event-based method: introduction of spatially configured stochastic reaction chambers (SCSRC). Math Biosci 2009;217:43–52.

[18] Rivière B, Epshteyn Y, Swigon D, Vodovotz Y. A simple mathematical model of signaling resulting from the binding of lipopolysaccharide with Toll-like receptor 4 demonstrates inherent preconditioning behavior. Math Biosci 2009;217:19–26.

[19] Nieman K, Brown D, Sarkar J, Kubiak B, Ziraldo C, Vieau C, et al. A two-compartment mathematical model of endotoxin-induced inflammatory and physiologic alterations in swine. Crit Care Med 2012;40:1052–63.

[20] Dong X, Foteinou PT, Calvano SE, Lowry SF, Androulakis IP. Agent-based modeling of endotoxin-induced acute inflammatory response in human blood leukocytes. PLoS One 2010;5(2):e9249.

[21] Foteinou PT, Calvano SE, Lowry SF, Androulakis IP. Multiscale model for the assessment of autonomic dysfunction in human endotoxemia. Physiol Genomics 2010;42(1):5–19.

[22] Scheff JD, Calvano SE, Lowry SF, Androulakis IP. Modeling the influence of circadian rhythms on the acute inflammatory response. J Theor Biol 2010;264(3):1068–76.

[23] Foteinou PT, Calvano SE, Lowry SF, Androulakis IP. A physiological model for autonomic heart rate regulation in human endotoxemia. Shock 2011;35:229–39.

[24] Yang Q, Calvano SE, Lowry SF, Androulakis IP. A dual negative regulation model of Toll-like receptor 4 signaling for endotoxin preconditioning in human endotoxemia. Math Biosci 2011;232(2):151–63.

[25] Lagoa CE, Bartels J, Baratt A, Tseng G, Clermont G, Fink MP, et al. The role of initial trauma in the host's response to injury and hemorrhage: Insights from a comparison of mathematical simulations and hepatic transcriptomic analysis. Shock 2006;26:592–600.

[26] Torres A, Bentley T, Bartels J, Sarkar J, Barclay D, Namas R, et al. Mathematical modeling of post-hemorrhage inflammation in mice: studies using a novel, computer-controlled, closed-loop hemorrhage apparatus. Shock 2009;32:172–8.

[27] Arciero J, Rubin J, Upperman J, Vodovotz Y, Ermentrout GB. Using a mathematical model to analyze the role of probiotics and inflammation in necrotizing enterocolitis. PLoS One 2010;5:e10066.

[28] Kim M, Christley S, Alverdy JC, Liu D, An G. Immature oxidative stress management as a unifying principle in the pathogenesis of necrotizing enterocolitis: insights from an agent-based model. Surg Infect (Larchmt) 2012;13(1):18–32.

[29] Arciero J, Ermentrout GB, Siggers R, Afrazi A, Hackam D, Vodovotz Y, et al. Modeling the interactions of bacteria and Toll-like receptor-mediated inflammation in necrotizing enterocolitis. J Theor Biol 2013;321:83–99.

[30] Barber J, Tronzo M, Harold HC, Clermont G, Upperman J, Vodovotz Y, et al. A three-dimensional mathematical and computational model of necrotizing enterocolitis. J Theor Biol 2013;322:17–32.

[31] Edelstein-Keshet L. Mathematical models in biology. New York, NY: Random House; 1988.

[32] Clermont G. Translational equation-based modeling Vodovotz Y, An G, editors. Complex systems and computational approaches to acute inflammation. New York, NY: Springer; 2013.

[33] Namas R, Ghuma A, Hermus L, Zamora R, Okonkwo DO, Billiar TR, et al. The acute inflammatory response in trauma/hemorrhage and traumatic brain injury: current state and emerging prospects. Libyan J Med 2009;4:97–103.

[34] Price I, Ermentrout B, Zamora R, Wang B, Azhar N, Mi Q, et al. *In vivo*, *in vitro*, and *in silico* studies suggest a conserved immune module that may regulate malaria parasite transmission from mammals to mosquitoes. J Theor Biol 2013;334:173–86.

[35] Zamora R, Azhar N, Namas R, Metukuri MR, Clermont T, Gladstone C, et al. Identification of a novel pathway of TGF-beta1 regulation by extracellular NAD+ in mouse macrophages: *in vitro* and *in silico* studies. J Biol Chem 2012;287:31003–31014.

[36] Kitano H. Systems biology: a brief overview. Science 2002;295(5560):1662–4.

[37] Angeli D, Ferrell J, Sontag ED. Detection of multistability, bifurcations, and hysteresis in a large class of biological positive-feedback systems. Proc Natl Acad Sci USA 2004;101:1822–7.

[38] Bagci EZ, Vodovotz Y, Billiar TR, Ermentrout GB, Bahar I. Bistability in apoptosis: roles of Bax, Bcl-2 and mitochondrial permeability transition pores. Biophys J 2006;90:1546–59.

[39] Marino S, Hogue IB, Ray CJ, Kirschner DE. A methodology for performing global uncertainty and sensitivity analysis in systems biology. J Theor Biol 2008;254(1):178–96.

[40] Levi MI, Smirnova OA, Stepanova NV. [Mathematical models of the primary immunological reaction (a review)]. Zh Mikrobiol Epidemiol Immunobiol 1974;11:113–20.

[41] Alt W, Lauffenburger DA. Transient behavior of a chemotaxis system modelling certain types of tissue inflammation. J Math Biol 1987;24(6):691–722.

[42] Nelson GW, Perelson AS. A mechanism of immune escape by slow-replicating HIV strains. J Acquir Immune Defic Syndr 1992;5(1):82–93.

[43] An G, Hunt CA, Clermont G, Neugebauer E, Vodovotz Y. Challenges and rewards on the road to translational systems biology in acute illness: four case reports from interdisciplinary teams. J Crit Care 2007;22:169–75.

[44] Natanson C, Hoffman WD, Suffredini AF, Eichacker PQ, Danner RL. Selected treatment strategies for septic shock based on proposed mechanisms of pathogenesis. Ann Intern Med 1994;120:771–83.

[45] Angus DC. The search for effective therapy for sepsis: back to the drawing board? JAMA 2011;306(23):2614–5.

[46] Nathan C. Points of control in inflammation. Nature 2002;420(6917):846–52.

[47] Freeman BD, Natanson C. Anti-inflammatory therapies in sepsis and septic shock. Expert Opin Investig Drugs 2000;9(7):1651–63.

[48] Opal SM, DePalo VA. Anti-inflammatory cytokines. Chest 2000;117(4):1162–72.

[49] Vodovotz Y, Constantine G, Rubin J, Csete M, Voit EO, An G. Mechanistic simulations of inflammation: current state and future prospects. Math Biosci 2009;217:1–10.

[50] Vodovotz Y. At the interface between acute and chronic inflammation: insights from computational modeling Roy S, Sen C, editors. Chronic inflammation: nutritional and therapeutic interventions. Florence, KY: Taylor & Francis; 2013.

[51] Namas RA, Bartels J, Hoffman R, Barclay D, Billiar TR, Zamora R, et al. Combined *in silico*, *in vivo*, and *in vitro* studies shed insights into the acute inflammatory response in middle-aged mice. PLoS One 2013;8:e67419.

[52] Mi Q, Rivière B, Clermont G, Steed DL, Vodovotz Y. Agent-based model of inflammation and wound healing: insights into diabetic foot ulcer pathology and the role of transforming growth factor-β1. Wound Repair Regen 2007;15:617–82.

[53] Li NYK, Verdolini K, Clermont G, Mi Q, Hebda PA, Vodovotz Y. A patient-specific *in silico* model of inflammation and healing tested in acute vocal fold injury. PLoS One 2008;3:e2789.

[54] An G, Mi Q, Dutta-Moscato J, Solovyev A, Vodovotz Y. Agent-based models in translational systems biology. WIRES 2009;1:159–71.

[55] Li NYK, Vodovotz Y, Hebda PA, Verdolini K. Biosimulation of inflammation and healing in surgically injured vocal folds. Ann Otol Rhinol Laryngol 2010;119:412–23.

[56] Li NYK, Vodovotz Y, Kim KH, Mi Q, Hebda PA, Verdolini Abbott K. Biosimulation of acute phonotrauma: an extended model. Laryngoscope 2011;121:2418–28.

[57] Brown BN, Price IM, Toapanta FR, Dealmeida DR, Wiley CA, Ross TM, et al. An agent-based model of inflammation and fibrosis following particulate exposure in the lung. Math Biosci 2011;231:186–96.

[58] Solovyev A, Mi Q, Tzen Y-T, Brienza D, Vodovotz Y. Hybrid equation-/agent-based model of ischemia-induced hyperemia and pressure ulcer formation predicts greater propensity to ulcerate in subjects with spinal cord injury. PLoS Comput Biol 2013;9:e1003070.

[59] Dutta-Moscato J, Solovyev A, Mi Q, Nishikawa T, Soto-Gutierrez A, Fox IJ, et al. A multiscale agent-based *in silico* model of liver fibrosis progression. Front Bioeng Biotechnol 2014;2:1–10. (Article 18).

[60] Kox WJ, Volk T, Kox SN, Volk HD. Immunomodulatory therapies in sepsis. Intensive Care Med 2000;26(Suppl. 1):S124–8.

[61] Marshall JC. Clinical trials of mediator-directed therapy in sepsis: what have we learned? Intensive Care Med 2000;26(Suppl. 1):S75–83.

[62] Angus DC, Linde-Zwirble WT, Lidicker J, Clermont G, Carcillo J, Pinsky MR. Epidemiology of severe sepsis in the United States: analysis of incidence, outcome, and associated costs of care. Crit Care Med 2001;29(7):1303–10.

[63] Reinhart K, Karzai W. Anti-tumor necrosis factor therapy in sepsis: update on clinical trials and lessons learned. Crit Care Med 2001;29 (7 Suppl.):S121–5.

[64] Wenner KA, Kenner JR. Anthrax. Dermatol Clin 2004;22(3):247–56. v.

[65] Turnbull PC, Leppla SH, Broster MG, Quinn CP, Melling J. Antibodies to anthrax toxin in humans and guinea pigs and their relevance to protective immunity. Med Microbiol Immunol (Berl) 1988;177(5):293–303.

[66] Iacono-Connors LC, Welkos SL, Ivins BE, Dalrymple JM. Protection against anthrax with recombinant virus-expressed protective antigen in experimental animals. Infect Immun 1991;59(6):1961–5.

[67] Ivins BE, Welkos SL, Little SF, Crumrine MH, Nelson GO. Immunization against anthrax with *Bacillus anthracis* protective antigen combined with adjuvants. Infect Immun 1992;60(2):662–8.

[68] Barnard JP, Friedlander AM. Vaccination against anthrax with attenuated recombinant strains of *Bacillus anthracis* that produce protective antigen. Infect Immun 1999;67(2):562–7.

[69] Brossier F, Weber-Levy M, Mock M, Sirard JC. Protective antigen-mediated antibody response against a heterologous protein produced *in vivo* by *Bacillus anthracis*. Infect Immun 2000;68(10):5731–4.

[70] Friedlander AM, Welkos SL, Ivins BE. Anthrax vaccines. Curr Top Microbiol Immunol 2002;271:33–60.

[71] Csete ME, Doyle JC. Reverse engineering of biological complexity. Science 2002;295(5560):1664–9.

[72] Kurata H, El-Samad H, Iwasaki R, Ohtake H, Doyle JC, Grigorova I, et al. Module-based analysis of robustness tradeoffs in the heat shock response system. PLoS Comput Biol 2006;2(7):e59.

[73] An G, Nieman G, Vodovotz Y. Computational and systems biology in trauma and sepsis: current state and future perspectives. Int J Burns Trauma 2012;2:1–10.

IV. TOOLS AND IMPLEMENTATION OF TRANSLATIONAL SYSTEMS BIOLOGY: THIS IS HOW WE DO IT

[74] Cavaillon JM. The nonspecific nature of endotoxin tolerance. Trends Microbiol 1995;3(8):320–4.

[75] Cavaillon JM, Adrie C, Fitting C, Adib-Conquy M. Endotoxin tolerance: is there a clinical relevance? J Endotoxin Res 2003;9(2):101–7.

[76] Eng RH, Smith SM, Fan-Havard P, Ogbara T. Effect of antibiotics on endotoxin release from gram-negative bacteria. Diagn Microbiol Infect Dis 1993;16(3):185–9.

[77] Dani C, Biadaioli R, Bertini G, Martelli E, Rubaltelli FF. Probiotics feeding in prevention of urinary tract infection, bacterial sepsis and necrotizing enterocolitis in preterm infants. A prospective double-blind study. Biol Neonate 2002;82(2):103–8.

[78] Land MH, Rouster-Stevens K, Woods CR, Cannon ML, Cnota J, Shetty AK. Lactobacillus sepsis associated with probiotic therapy. Pediatrics 2005;115(1):178–81.

[79] Marshall JC, Deitch E, Moldawer LL, Opal S, Redl H, Poll TV. Preclinical models of shock and sepsis: what can they tell us? Shock 2005;24(Suppl. 1):1–6.

[80] Mathew S, Bartels J, Banerjee I, Vodovotz Y. Global sensitivity analysis of a mathematical model of acute inflammation identifies nonlinear dependence of cumulative tissue damage on host interleukin-6 responses. J Theor Biol 2014;358:132–48.

[81] Namas R, Namas R, Lagoa C, Barclay D, Mi Q, Zamora R, et al. Hemoadsorption reprograms inflammation in experimental gram-negative septic fibrin peritonitis: insights from *in vivo* and *in silico* studies. Mol Med 2012;18:1366–74.

[82] Peitzman AB, Billiar TR, Harbrecht BG, Kelly E, Udekwu AO, Simmons RL. Hemorrhagic shock. Curr Probl Surg 1995;32(11):925–1002.

[83] Chaudry IH, Ayala A, Ertel W, Stephan RN. Hemorrhage and resuscitation: immunological aspects. Am J Physiol 1990;259(4 Pt 2):R663–78.

[84] Chaudry IH, Ayala A. Mechanism of increased susceptibility to infection following hemorrhage. Am J Surg 1993;165(2A Suppl.):59S–67S.

[85] Galmarini S, Bianconi R, Klug W, Mikkelsen T, Addis R, Andronopoulos S, et al. Can the confidence in long range atmospheric transport models be increased? The pan-European experience of ensemble. Radiat Prot Dosimetry 2004;109(1–2):19–24.

[86] Palmer TN, Raisanen J. Quantifying the risk of extreme seasonal precipitation events in a changing climate. Nature 2002;415(6871):512–4.

[87] Diffenbaugh NS, Field CB. Changes in ecologically critical terrestrial climate conditions. Science 2013;341(6145):486–92.

[88] Schewe J, Heinke J, Gerten D, Haddeland I, Arnell NW, Clark DB, et al. Multimodel assessment of water scarcity under climate change. Proc Natl Acad Sci USA 2014;111(9):3245–50.

[89] Prudhomme C, Giuntoli I, Robinson EL, Clark DB, Arnell NW, Dankers R, et al. Hydrological droughts in the 21st century, hotspots and uncertainties from a global multimodel ensemble experiment. Proc Natl Acad Sci USA 2014;111(9):3262–7.

[90] Gneiting T, Raftery AE. Atmospheric science: weather forecasting with ensemble methods. Science 2005;310(5746):248–9.

[91] Thyagarajan T, Totey S, Danton MJ, Kulkarni AB. Genetically altered mouse models: the good, the bad, and the ugly. Crit Rev Oral Biol Med 2003;14(3):154–74.

[92] Osuchowski MF, Remick DG, Lederer JA, Lang CH, Aasen AO, Aibiki M, et al. Abandon the mouse research ship? Not just yet! Shock 2014;41(6):463–75.

[93] An G, Bartels J, Vodovotz Y. *In silico* augmentation of the drug development pipeline: examples from the study of acute inflammation. Drug Dev Res 2011;72:1–14.

[94] Ferrucci L, Ble A, Bandinelli S, Lauretani F, Suthers K, Guralnik JM. A flame burning within. Aging Clin Exp Res 2004;16(3):240–3.

[95] Plackett TP, Boehmer ED, Faunce DE, Kovacs EJ. Aging and innate immune cells. J Leukoc Biol 2004;76(2):291–9.

[96] Franceschi C, Bonafe M, Valensin S, Olivieri F, De LM, Ottaviani E, et al. Inflamm-aging. An evolutionary perspective on immunosenescence. Ann NY Acad Sci 2000;908:244–54.

[97] Namas R, Ghuma A, Torres A, Polanco P, Gomez H, Barclay D, et al. An adequately robust early TNF-α response is a hallmark of survival following trauma/hemorrhage. PLoS One 2009;4(12):e8406.

[98] Scheff JD, Mavroudis PD, Foteinou PT, An G, Calvano SE, Doyle J, et al. A multiscale modeling approach to inflammation: a case study in human endotoxemia. Shock 2013;244:279–89.

4.4

Agent-Based Modeling and Translational Systems Biology: An Evolution in Parallel

THINGS DOING THINGS AND THE WISDOM OF CROWDS

We have made the case throughout this book of the power of mathematical representation as a means of characterizing the hidden order behind the varied observations we make in the natural world. The prior sections on data-driven modeling and the use of mathematical equations to represent biological processes provided examples of the relatively direct application of mathematical techniques to the biomedical Scientific Cycle; these approaches directly harken back to the genius of Newton and the elegant mapping between differential calculus and classical mechanics. However, we have also seen through our prior discussions that the field of biology has certain characteristics that make it difficult in certain circumstances, if not impossible, to utilize traditional differential equation modeling to represent mechanism. A recurrent theme throughout this book is the need to find the correct tool for a particular task, and this requires a return to an examination of the basic assumptions inherent to a particular method or approach.

We apply this process to the use of differential equations for biomedical knowledge representation by examining the assumptions required when they are used to represent biological systems. Most notable of these assumptions in the case of ordinary differential equations (ODEs) is that the system being studied needs to be characterized, as some level, as a well-mixed container of multiple types of homogeneous components. Note that this requirement holds true even for multicompartment ODEs, where each individual compartment, however finely it is divided, needs to be treated as a homogeneously well-mixed system; we will see shortly the consequences that result if this subdivision process is continued nearly *ad infinitum*. The need to assume a well-mixed system means that the variables used by the ODE to represent properties and characteristics of the system are single values for an aggregated population metric assumed to be the same across the entire population of components. Therefore, if some of the effects of the interactions within that population are affected by inhomogeneity within a particular component type, such as their location or orientation, then the assumptions made in constructing the ODE may not be valid. The most common circumstance that leads to the breaking of the well-mixed system assumption is when the spatial configuration of the components making up the particular system matters. Such is often the case when looking at biological systems, particularly in terms of examining the behavior of multicellular organisms. For these organisms (such as humans) discrete organ systems, which are themselves made up of different populations of cells in specifically defined configurations and architectures, which then interact to produce the behaviors seen at the tissue and organ levels. A related, practical example within the realm of Translational Systems Biology revolves around the fact that some clinically-relevant data are inherently spatial; examples include tissue biopsies (histology), wounds on the skin, and rashes. Approximating these complex spatial patterns with ODEs is difficult.

While many aspects of biomedical systems can be characterized effectively by making abstracting assumptions regarding the lack of importance of the spatial configurations of biological tissue, there is a point at which the accumulating assumptions required for ODE representation of a particular problem causes a breakdown in the mapping between the mathematical representation and the system it purports to represent. Sometimes such spatial issues

111

can be managed with partial differential equations (PDEs), which add variables that define spatial positioning and orientation to the time dimension of ODEs, and indeed PDE approaches are also heavily employed in biomedical models. But, as with all continuous, equation-based modeling methods, the application of PDEs requires the translation of biological entities into representative variables that, to some degree, still reflect an aggregated population assumption (now discretized not only by time but also of space). This does not present an intractable issue, but can sometimes provide an additional potential hurdle to our goal of facilitating the dynamic knowledge representation of biological systems. Therefore, it behooves us to investigate whether there are additional computational modeling methods that might more readily map to what we recognize as properties of biological systems, one of the most basic of which is that they result from the interactions among multiple populations of cells. In the last few decades of the twentieth century just such a method was developed, with roots in computer science as opposed to mathematics, which seemed to map particularly well to the study of biology. This method is called agent-based modeling.

Agent-based modeling is an object-oriented, discrete-event, rule-based method for creating computer simulations [1–5]. An agent-based model (ABM) represents a dynamic system as a population of interacting semiautonomous components or *agents*; as such developing and using an ABM requires thinking of a system as a set of *objects*. These agents are semiautonomous in that, while they are defined by specific behavioral *rules*, they execute those rules based on their own local information, i.e., their behavior is not controlled or dictated by a global entity. This process involves the agents examining their local neighborhood and then executing their rules based on what they find; i.e., acting through a series of *discrete events*. Population effects arising from the interactions between individual agents produce overall system behavior, which is often characterized using a different scale of observable metrics than the variables used to determine agent behavior. Since a system can be decomposed at multiple different levels of components, development of an ABM requires that selection with respect to the primary component level to be utilized in the representation of the system. This component level becomes the simulation agent level of the ABM; for ABMs used to study biological systems a common such selection is the choice of using cells as the primary agent level. An ABM *agent class* provides a grouping of similar types of agents together, defined by specific properties and rules governing their identity and behavior. An ABM creates a population of individual agents from each agent class, where each individual agent possesses the behavioral rules and defined properties of its agent class. However, once created and running in an ABM the individual agents can have diverging behavioral paths based on differences in their local conditions. In short, ABMs can be simply be thought of viewing a system as arising from things doing things.

Agent-based modeling is actually a very intuitive way of looking at systems, and is particularly well suited to biology, which can readily be seen as different populations of interacting cells. The intuitive nature of agent-based modeling is further enhanced by the fact that ABM rules are often expressed as conditional statements ("if-then" statements). This makes an ABM particularly effective in expressing the mechanisms identified through the course of basic biomedical research; i.e., the presence of a ligand leads to the binding of a cellular receptor, which leads to a particular cellular behavior. However, it should be noted that the general conditional nature of simulation agent rules does not preclude the use of other types of mathematical or computational models (i.e., differential equation, stochastic or dynamic network) to represent agent rules [6–8]. Here, we see a convergence between ODE approaches and agent-based modeling. In our example of a successively compartmentalized ODE, the question that arises from that process is: "how do the compartments interact?" Are there multiple compartments present? Do they all act the same? Attempting to answer these questions, while keeping in mind the nature of biological things, causes one to realize that a multiple compartmentalized ODE model eventually results in an ABM.

Additionally, regardless of how the agent rules are implemented, ABMs offer the ability to achieve a close mapping between the natural language expression of mechanisms present in publications (the current means by which this knowledge is communicated within the community), and the structure of ABM [9,10]. The ease of this mapping facilitates the use of agent-based modeling as a means of dynamic knowledge representation that can be used to integrate multiple mechanisms and hypotheses, particularly for nonmathematicians/computational scientists, such as the general biomedical research population [9–12]. The key point to be noted here is that ABMs are based on *knowledge* as opposed to data, with our distinction that *data* represents the observations made and *knowledge* represents the mechanistic interpretation of those observations. Here, we harken back to the Scientific Cycle and the difference between the processes of induction and abduction that follow from empirical observations. As knowledge-based models constructed by implementing hypothesized (or, more precisely, "best guesses at") bottom-up mechanisms, ABMs address different scientific questions than traditional, data-driven equation-based inductive models. Very often, equations are used to attempt to explain data through an inductive process; i.e., what set of equations has previously been successful at fitting to similar data? We have described variations of this process in Chapters 4.2 and 4.3, and demonstrated how they can contribute to the goals of Translational Systems Biology. However, the generation of a model directly from data cannot be done with ABMs, and because of this property ABMs should not be considered a form of "data analysis." ABMs cannot be developed directly from a mass of raw data, but rather are dynamic representations of mechanistic knowledge that can be used to generate simulated data/behavior. This places an ABM very much in the world of experimental objects, and allows them to be used to examine and potentially falsify the assumptions embedded in them. The behavioral output of the ABM under different conditions can be compared to real-world data sets in an ongoing process of validation (or, more correctly, successive attempts at Popperian falsification) intended to increase trust in the model.

As we have noted, agent-based modeling, because of its emphasis on "things doing things," is often more intuitive for biomedical researchers than more formal mathematical methods, making for a clearer mapping between the biology and the computational representation, and providing a lower threshold barrier for researchers to integrate *in silico* methods with traditional *in vitro* and *in vivo* experiments [12]. This ability to more intuitively connect biologists with computational models is in keeping with the central goals of Translational Systems Biology: facilitating the acceleration of the overall iterative Scientific Cycle. We can see the framework for a process where inductive models are applied to large data sets, wet lab experiments are carried out to investigate the mechanisms inferred from the inductive model, and the experimentally confirmed mechanisms are used as the basis of ABMs that would close the discovery loop by attempting to recapitulate the original data set [13,14]. We will see the natural extension of this concept at a larger scale in Chapter 4.5, but before we proceed to that point let us examine the nature of ABMs in more detail, and the specific role of agent-based modeling in the development of Translational Systems Biology.

Before we delve into these topics, we will, as we have throughout this book, take a step back and make sure we address any recognized barriers arising from Bacon's Idols of the Mind that could hamper our progress. Here, we readily see an example of the Idols of the Marketplace, since there are potential linguistic pitfalls that can cause confusion. In this particular case, it is the difference between the terms *agent-based modeling* versus *agent-directed simulation/multiagent system*. This is an example of how the ambiguity associated with natural language, that common terms can take on different meanings with just enough relatedness as to maximize confusion. It should be noted that the following distinction is by no means intended to be a definitive description of the distinction between these two entities; rather, it is intended to clarify the differential usage of the term "agent" within the context of our discussion; this exercise can be thought of us setting the rules before playing Wittgenstein's language game. In our game, *agent-based modeling* is a simulation method: the model being developed and used is explicitly trying to mimic some set of qualities and attributes of a specifically targeted system to be studied. Within the context of Translational Systems Biology, that targeted system is almost invariably a biological object (organism, tissue, or cell). The simulation agents making up the ABM are mapped to some real-world objects, and selected characteristics of the real-world object are reflected in the nature of the rules embedded within the simulation agent. As we have noted, the main advantage of agent-based modeling is the ability to represent populations of real-world objects at the individual level with simulation agents; therefore, in most cases ABMs consist of many individual simulation agents derived from a single agent class. Alternatively, an agent-directed simulation or multiagent system refers primarily to a computer/software architecture solution in which computer agents perform tasks related to the implementation of a particular computing goal. Often the computer agents have some "intelligent" decision-making capacity that allows them to manage the information flow within a particular software implementation. In general, there is not a specific, real-world reference object for a computational agent in a multiagent system, rather there is a set of recognized tasks in information flow management that can be expressed as a set of algorithms and packaged for execution. For purposes of our discussion about Translational Systems Biology, the term *simulation agent* will be used in an agent-based modeling context and refer to simulations of real-world biological objects, whereas the term *computational agent* will be used in an agent-directed or multiagent system context and refer to a set of encapsulated information flow managing algorithms and decision processes.

PROPERTIES OF AGENT-BASED MODELS

Part of the difficulty with the terminology associated with agent-based modeling is that, as opposed to many other types of computational modeling methods, ABMs do not have a formal technical specification and definition. As we started off our discussion of agent-based modeling with a set of adjectives ("object-oriented," "rule-based," "discrete-event"), ABMs are primarily described by a list of attributes as opposed to a formally described structure. Interestingly, this is similar to how biological organisms have been traditionally categorized. This sense of "fuzziness" about agent-based modeling had, for many years, limited their use in the quantitative disciplines, where these models were treated more as software objects than a formal modeling methodology. To some degree, this remains true: ABMs still do not have a formal specification and are not generally able to be subjected to formal mathematical analysis in terms of characterizing their behavior (Note: However, as computer programs, ABMs can be subjected to formal computer program assessment methods such as model checking and code verification. These methods determine if the program is performing and behaving as it is expected/designed to do but cannot make statements about the type of behavior they are intended to examine). On the other hand, much of their expressive power comes from the fact that ABMs are not particularly constrained in their formal structure, allowing modelers to more readily represent their concepts as they intend to rather than morphing them to fit the requirements of a particular specification (more on this later, as well). Recognizing the challenges associated with defining ABMs, we will try to establish their relationship to other types of modeling methods that might be encountered in the context of biomedical research. ABMs are related to other spatially discrete modeling methods, most notably cellular automata, though the mobile capability ABM agents and ability to represent a wider range of model topologies could lead one to think of cellular automata as a special type of limited ABM. In a similar fashion, cellular Potts models can also be considered a special type of ABM [15]. Also, neural nets can be considered ABMs, with the nodes representing members of an agent class of "nodes," and the network structure being the model topology. In practice, most ABMs have several characteristics that set them apart from other object-oriented, rule-based modeling systems (such as Petri nets, Boolean, or Bayesian Networks), even though at its purest definition, all these approaches could all be potentially viewed as and implemented as ABMs. Here we provide a list of some of the primary attributes of an ABM:

1. ABMs facilitate *spatial representation*. In an ABM, agent behavior is driven by interactions determined by an agent's interaction neighborhood, which defines what the agent considers its local environment with which it can sense and communicate. Some examples of agent neighborhood configurations are: a two-dimensional square grid (very common), a three-dimensional cubic space [6,11], two- or three-dimensional hexagonal space [10,16] or as a network topology [17]. This definition of an agent neighborhood is consistent with the bounded nature of the sense-and-respond, message passing communication and interaction capabilities of biological objects.

2. ABMs utilize *parallelism*. In most cases, an ABM incorporates many individual member agents from a particular agent class, forming a population of agents, each capable of having different behavioral paths. These multiple, different behavioral paths produce population dynamics that are the observable, system-level output of the ABM. A classic example of this phenomenon is the behavior of flocks of birds, in which simulations utilizing relatively simple interaction rules among birds can lead to sophisticated flocking patterns without an overall controller [18]. There are some exceptions to this case, particularly when ABMs are nested together; there only be one or a few resulting "superagents" when this approach is utilized (more on this below). But for the most part, ABMs utilize the power of large populations of interacting components to generate their behavior.

3. ABMs incorporate *stochasticity*. Many biological systems have behaviors that appear to be random [19,20]. Probabilities of a particular behavior can be determined for the population as a whole, in many ways this is exactly the raw output from a basic science experiment. That information can be used to generate a probability function for the behavior of a single agent that is then incorporated into the agent's rules. As a population of agents execute their rules during the course of a simulation, each agent follows a particular behavioral path as its behavior rules' probabilities are resolved as the simulation progresses. A set of multiple output paths is thusly generated from a single ABM, producing system-level behavioral state spaces representing the set of population-level biological observations.

4. ABMs are intrinsically *multiscale*. The mandatory structure of even the simplest ABM incorporates three scales of system organization. The first scale of system representation is the knowledge that goes into creating the behavioral rules (Scale #1). The second scale of system representation is that level of the agents that execute those rules (Scale #2). Finally, the third scale is the population-level effects that result from the aggregated agent interactions (Scale #3). As seen with our prior discussion on biocomplexity, this multiscale structure allows ABMs to address exactly those aspects of the organization of biological systems that challenge the intuitive extrapolation of mechanisms from molecular, to cell, to organ, to organism. Additionally, these organizational levels can theoretically be *nested*, to provide a comprehensive depiction of a multiscale biological system, making ABMs well suited for creating modular models.

5. ABMs can be constructed readily using *incomplete and abstracted knowledge*. When constructing an ABM, it is advantageous at the outset to keep the rules as simple as possible, even at the expense of some detail; this is a general modeling practice used for dynamic knowledge representation that we discussed in Chapter 3.2.

Therefore, meta-analyses of existing basic research often guide the development of an ABM [21]. In keeping with the concept of useful and iterative dynamic knowledge representation, ABMs constructed with incomplete and abstracted mechanisms can provide qualitative conceptual verification of the hypotheses they instantiate [22]. An iterative process of refinement of an ABM will lead to increased detail and a greater fidelity of the mapping between the ABM and its biological counterpart. This enhances the correlation between simulation results and the real-world behaviors and lead to a greater confidence in the ability of the ABM to describe observable phenomena.

The combination of this list of attributes leads to the central functional hallmark of ABMs: they are able to generate system-level behaviors that could not have been reasonably inferred from, and often may be counterintuitive to, the underlying rules of the agents alone. ABMs are able to generate this type of behavior due to the locally constrained and stochastic nature of agent rules, and the population effects of their aggregated parallel interactions. For example, in the bird flock example described above, an initial observation would suggest an overall leader, thereby requiring a means of determining rules for flock-wide command and control communication. This, however, is not the actual case; birds function on a series of locally constrained, neighborhood-defined interaction rules, and the flocking behavior emerges from the aggregate of these interactions [18]. The capacity to generate nonintuitive behavior is a vital advantage of using ABM for conceptual model verification, as often the translation of generative mechanisms to system-level behavior produces paradoxical and unanticipated results that break a conceptual model.

Agent-based modeling was pioneered in the areas of ecology, social science, and economics, but in the last decade this class of model has been increasingly used in the biomedical arena to study sepsis [10,11,23,24], cancer [6,16,25–27], cellular trafficking [28–32], wound healing [33–35], as well as intracellular processes and signaling [1,7,36–41]. Because of the very intuitive way of viewing biological systems as groups of interacting cells, many biomedical ABMs focus on cells as the primary simulation agent level. In terms of mapping to biological knowledge, cells form an easily recognized level of "encapsulated complexity," with the dual advantages that they are highly studied as a unit (i.e., cellular biology), and are readily amenable to being treated with input–output rules [5]. Furthermore, as we have noted above, ABM agent rules can at some level be considered analogous to the equations that make up a mathematical model, with multiple examples of embedding complex mathematical models within a cell-level ABM agent [5–8,35,42,43]. These examples emphasize the potential unifying role of agent-based modeling as a means of "wrapping" different simulation methodologies, as well as allowing ABMs to potentially incorporate the detailed pathway models generated in classical systems biology. We will explore this concept of using the metastructure of ABMs as an integrating framework for mechanistic biomedical knowledge later in this chapter [14].

AGENT-BASED MODELING OF INFLAMMATION AND THE DEVELOPMENT OF TRANSLATIONAL SYSTEMS BIOLOGY

The use of agent-based modeling in the biomedical research community has increased dramatically since the year 2000, and is now a generally accepted means of performing computational biology. We were among the earliest proponents of agent-based modeling in the biomedical community, and the utilization of ABMs was an integral part in our development of Translational Systems Biology [44,45]. We will present the history of the utilization of ABMs as part of the development and evolution of the goals and tenets of Translational Systems Biology. In addition to providing perspective into the foundational aspects of Translational Systems Biology, this survey of our ABMs will demonstrate the conceptual differences between how we have viewed and used ABMs and how other research groups have utilized ABMs in a more traditional biomedical research fashion. The key point we hope to communicate is that in formulating the concept of Translational Systems Biology, our starting motivation for computational modeling was, and remains, to target clinical conditions in order to improve the clinical implementation of biological knowledge by bridging the gulf between basic biomedical research and the bedside. As such, our modeling efforts, whether using equation- or agent-based modeling, first focused on representation of clinical phenomena as opposed to targeting the more traditional reductionist goal (be it experimental or computational) of obtaining detailed mechanistic insights. In short, our approach was driven by an explicit attempt to integrate and synthesize existing knowledge to identify whether clinical phenomena could be reproduced, and then be used as the jumping off point for additional investigation. The pursuit of basic biological insight was secondary to our emphasis on generating clinically relevant simulations, though, we reasoned that these insights would be derived in the process of adding mechanistic details to models that could predict the behavior of cells, tissues, organs, whole organisms, and populations. In fact, we will see how Translational Systems Biology, through its evolution, does indeed incorporate these reductionist methods (after all, reductionism remains the only way to identify mechanisms) in the service of our clinically relevant goal. However, it is the primacy of the clinical emphasis that sets Translational Systems Biology apart from other approaches through its role to bridge basic mechanisms and clinical/epidemiological science.

As we have noted previously, our focus on the inflammatory response as the target for our initial modeling efforts, including agent-based modeling, was an outgrowth of our clinical and experimental background in the treatment and study of sepsis and trauma. However, we believe this applied focus was fortuitous, because inflammation, as we have seen, is one of the most basic and ubiquitous processes in biology, and in addition to growth, metabolism, and replication, the response to injury leading into repair is a core function of all organisms. As such, our initial work on inflammation afforded us the opportunity to expand the scope of our investigations to a degree of near universal applicability, both in terms of biology (given the pervasive role of inflammation in a whole host of pathophysiological processes) as well as research process (i.e., all research on "systems" diseases similarly faced the Translational Dilemma). Finally, our use of multiple modeling tools, including data-driven models (Chapter 4.2) and equation-based models (Chapter 4.3) in addition to agent-based modeling, to study inflammation allowed us to cross-check assumptions and compare simulation methods and outcomes. This means of using different modeling methods to arrive at the same behaviors and conclusions is termed *cross-platform validation*. A central principle of our approach centered around that hypothesis that core mechanisms of inflammation are conserved and can therefore be observed at multiple levels, including both mean concentrations of mediators in biofluids (from equation-based models) and generated spatial patterns at the tissue level (from ABMs).

INITIAL SIMULATIONS OF CLINICAL POPULATIONS AND *IN SILICO* CLINICAL TRIALS

As we previously noted in Section 3, a primary driving tenet of Translational Systems Biology involves the use of dynamic computational modeling to bridge basic mechanisms to clinically identifiable scenarios, with an ultimate goal of producing *in silico* clinical trials. The fundamental place of this goal in the development of Translational Systems Biology is reflected in the fact that the first ABMs we produced (which were among the first cell-level ABMs published of any disease process) were aimed at generating simulated clinical populations at risk for sepsis [23,24,26]. The experience we gained in developing and using these models led to the realization that even abstract agent rules could produce very recognizable dynamics, which in turn could provide deep insights into the essential characterization of a disease process. We were focused specifically on deep insights with a directly translational application, as we discuss below.

The first of these ABMs, initially developed in 1999 and published in 2001, was one of the first examples of the use of agent-based modeling for biomedical dynamic knowledge representation [23]. This early ABM of the Systemic Inflammatory Response Syndrome (SIRS) viewed the inflammatory process as being governed by cellular interactions at the endothelial-blood interface, and was termed the Innate Immune Response (IIR) ABM. This abstraction reflected a prevailing interest at that time in the role of endothelial activation and its ability to propagate inflammation from a local to a systemic process as a driving process in sepsis. The IIR ABM played a unifying role in the study of systemic inflammatory disorders by demonstrating that the mechanistic basis of inflammation was the same whether the initiating insult was infectious, as in classical sepsis, or tissue damage, as in severe trauma. It focused on reproducing the early, pro-inflammatory phase of sepsis that at that time occupied (and to a great degree, continues to occupy) the bulk of basic and clinical research in the area. Because of the emphasis on the early events in the systemic inflammatory response it was this phase of that was targeted for pharmaceutical intervention. The simulations of the initial IIR ABM demonstrated formally for the first time that the degree of insult alone, either infectious or noninfectious, could be used to drive an intact and otherwise normally configured inflammatory system into a forward-feedback pro-inflammatory state.

The endothelial-surface IIR ABM was then extended by the incorporation of anti-inflammatory pathways [24], which allowed the ABM to now replicate all four primary outcomes from systemic insult. These four outcomes are: 1) healing, 2) the forward-feedback pro-inflammatory SIRS phenotype, 3) the negative-feedback immunoparalysis of multiple organ dysfunction syndrome/multiple organ failure (MODS/MOF) (previously poorly characterized and at that time, under-recognized in its clinical importance), and 4) Overwhelming Insult that lead to system death. The enhanced IIR ABM generated these four distinct clinical trajectories of model-system behavior purely by altering the degree of initial perturbation/insult. Selected levels of initial insult were used to generate simulated clinical populations matching the then 28-day mortality of the clinical sepsis population, and these simulated populations were then used to execute a series of what were among the first set of *in silico* trials published (along with our concurrent work using ODEs [46]). Remember, this was the era immediately following the failed anti-cytokine/mediator trials of sepsis, a series of outcomes that led to the recognition of the Translational Dilemma. The published pharmacologic properties of this series of mediator-targeting compounds were inputted into the ABM simulating a sepsis population being treated with experimental, anti-cytokine drugs. The goal of this exercise was to evaluate the conceptual soundness of these proposed interventions. Since these interventions represented the "best" of the drug development pipeline, the question we sought to ask and answer was: "Given our understanding and conceptual model of the sepsis response, does an intuitively sound intervention behave as we would expect it to?" Since our goal was to evaluate the soundness of the conceptual model, all interventions were assumed to perform exactly as they were intended to in terms of blocking their targeted mediators. The efficacies of these interventions were then evaluated against a simulated control population matching reported mortality rates for sepsis in a series of *in silico* clinical trials. Simulated trials were carried out for 28 days of simulated time, and overall mortality was used as the primary end point. These *in silico* clinical trials demonstrated that none of the mediator-directed interventions led to a statistically significant improvement in simulated patient outcome, including a set of hypothesized immune augmenting interventions and combination anti-cytokine therapy intended to overcome possible pathway redundancy. When these results were compared to retrospectively published clinical trials, this work provided one of the earliest examples of how dynamic knowledge representation could be used as a means of assessing the conceptual basis of a proposed intervention by visualizing the global consequences of intervening in a particular pathway, and determining whether it was actually a good idea to intervene at that point. This demonstration, that when simulated what appeared to be intuitively plausible points of mechanistic intervention, substantiated and vetted by the existing preclinical research pipeline, did *not* in fact behave as expected when placed in a systemic context, was the first inkling of the potential usefulness of agent-based modeling and dynamic knowledge representation for hypothesis verification. The results of this study led us to pursue the use of abstracted ABMs to reduce the set of plausible hypotheses and thereby help direct future investigation by eliminating therapeutic dead ends, a process consistent with Popperian falsification.

In addition to providing a template for future uses of ABMs to perform additional *in silico* clinical trials (see below), the ability to formalize the then current state-of-the-art mechanistic hypothesis allowed us to examine why these anti-cytokine therapies failed. The prevailing thought at the time of their development was to target the pro-inflammatory processes early and vigorously in order to stop the runaway train effect of excess inflammation in its tracks. As a result, the clinical trials of anti-inflammatory interventions attempted to give large, early doses to try and "jolt" the system back into equilibrium. When we simulated these clinical trials, since we were aiming to evaluate the soundness of their conceptual models, we assumed that the interventions did exactly what they were intended to do for the period of time the drugs were expected to work. We did this to avoid the potential issues with respect to bioavailability or potency that were raised by the primary investigators when those trials were found to have failed. What we found during the course of the simulations was that the interventions did indeed do exactly what they were designed to do: for their defined period of efficacy their respective targeted cytokines did disappear. However, in the absence of those cytokines, the damage present in the system, not being subjected to the responses intended to contain and suppress them, merely persisted through this phase and remained to retrigger the inflammatory response once the drugs wore off. So while it was true that the early pro-inflammatory accentuation was suppressed by the anti-cytokine interventions, so too was any recovery the system could have made during the period of the intervention, with the end result that when the efficacious period of the drugs had passed, the negative dynamics just resumed from that point on. This provided a critical insight that, at the time, was not generally recognized: that to a great degree, the dysfunction that generated sepsis and MOF may not have been due to too vigorous an inflammatory response, but rather an insufficient inflammatory response that could not contain and sufficiently resolve the initial insult. Importantly, this prediction corresponded well with those of ODE models of inflammation described in Chapter 4.3 [46,47]. Novel and unprecedented at that time, it is only in the subsequent decade that this phenomenon has become recognized as a vital intrinsic issue in the clinical courses seen in the Intensive Care Unit population [48].

PROVIDING NEW PERSPECTIVES ON CLINICAL CONDITIONS

As potentially powerful as these global systemic inflammation simulation experiments were in terms of predicting what types of interventions would fail, they unfortunately could not suggest interventions which could actually work. As such, these models could not completely fulfill our desired goal of providing a means for "useful failure." What became apparent was that the conclusions drawn from these simulations were, to a great degree, too broad and too far-reaching: realizing that attempting to block an endogenously generated mediator was a bad idea was a very important insight, but this insight did not provide a tangible strategy regarding a possibly successful therapeutic strategy targeted to a particular clinical manifestation of critical illness. It was apparent that given their level of abstraction, these ABMs did not have sufficiently detailed representations of the potentially relevant control structures of inflammation in order to apply a feasible engineering strategy to identify effective interventions. At one level, this could be seen as evidence that our strategy to utilize abstract models was misguided. However, we would counter with the fact that we are just following the good modeling practices suggested by the quotes from G.E.P. Box (*"All models are wrong, but some are useful."*) and A.N. Whitehead (*"The aim of science is to seek the simplest explanation of complex facts... Seek simplicity and distrust it."*): we have utilized a simplifying abstraction to generate a useful model, but now we need to proceed forward. The key question now would be: how to move forward in a rational and systematic fashion? We do not want to get caught up in the same chase of infinite regress that bedevils standard reductionist science as well as all too much systems biology work. Fortunately, we can turn to the clinical imperative intrinsic to Translational Systems Biology, and see that, in particular, agent-based modeling offers several avenues for looking at disease states with new, more integrative and synthesizing perspectives. This approach is completely consistent with our concept of dynamic knowledge representation, and here provides the opportunity to put together the multiple pieces of pathophysiology we know are acting in concert in the clinical setting, but are unable, given the constraints of experimental biology, to represent effectively using biological proxy models.

Seeing the Whole Elephant: Agent-Based Modeling of Multiple Organ Systems

We have already started off one of our chapters with the old story of the blind men of Hindustan, who all try and describe an elephant using each by feeling a separate portion of the elephant's body. The similarity between this condition and the fractured structure of the biomedical research community is too obvious and useful for us to not use the analogy again. At this point, however, we are prepared to provide a solution to the blind Hindustanis' dilemma, by taking advantage of the modular, multiscale nature of agent-based modeling.

Looking at our initial IIR ABMs, it is evident that one direction to move forward was to address our initial abstraction that the entire human body could be represented by a single endothelial-blood interface. The human body consists of multiple, interacting organs, and the clinical manifestation of disease likely results from dysfunction of those organs. Various cells and molecules can move from organ to organ via the circulation, which in turn consists of blood vessels that have a lining of endothelial cells. To increase the level of detail of our ABM, we sought to proceed using the unifying concepts generated from our initial set of ABMs, taking advantage of the intrinsic modular, multiscale nature of this modeling framework and the aforementioned endothelialized structure of these organs. Therefore, we proceeded with a structural/anatomic approach to multiscale modeling that utilized the modular property of ABMs to link individual organ ABMs in a multiscale, multiorgan architecture. The result of this refinement was an ABM of the gut-lung axis of systemic acute inflammation and multiple organ failure (MOF) [11]. This MOF ABM represented multiple structural and anatomic spaces as endothelial and epithelial surfaces aggregated by cell-type into organ-specific tissues, which in turn were configured to replicate organ-to-organ interconnections and cross talk, which is assumed to occur, at least in part, through a shared, endothelialized circulatory system. This is an explicit example of how ABMs manifest their intrinsic multiscale and modular properties. This integrating architecture also *translates* knowledge across domain specialties (molecular biology to clinical critical care), representing molecular and cellular mechanisms and behaviors derived from *in vitro* studies, extrapolated to *ex vivo* tissue experiments and observations, leading to patterns of organ-specific physiology, and finally simulating clinically relevant, interconnected, multiorgan physiology including the response to ventilator support of acute respiratory failure. In addition to now providing a clinically relevant output (i.e., respiratory failure and the need for organ support), the MOF ABM also allowed us to examine certain characteristics of the gut-derived pro-inflammatory compound that is circulated in the mesenteric lymph and induces pulmonary inflammation. Simulations with the MOF ABM we performed to integrate the time courses of a set of factors generated by the intestine following ischemic insults with the onset of pulmonary inflammation/failure. These simulations suggested that the mesenteric lymph inflammatory compound was not a primary inflammatory cytokine, nor a translocating luminal compound due to increased intestinal permeability, but rather a substance reflecting cellular damage of gut tissue with properties consistent with damage-associated molecular patterns (DAMPs). This last hypothesis remains to be confirmed completely by the critical care research community, but at this time appears to be consistent with ongoing research in this area [49]. Understanding the source of this MOF propagating compound can therefore direct investigations toward interventions aimed at preserving enterocyte integrity (as opposed to permeability alone), and focuses the molecular biology onto a different set of pathways than previously considered.

Characterizing the Continuums of Clinical Disease Space

In the course of characterizing the dynamics of the transition from health to disease, it became readily apparent that in the vast majority of cases there was not a single, isolated factor that defined that transition; rather, there was a constellation of factors that needed to align in order to disrupt the robust, baseline healthy condition. This fact is increasingly obvious today, as we see the effects of comorbidities (obesity, diabetes, cardiovascular disease, cancer, debilitating age) on the critically ill population and their responses (or more commonly, nonresponse) to inflammatory stimuli. Just as with all other aspects of biomedical research, so, too, has the identification of potential comorbid factors been deluged with an explosion of data. Here again is the danger of infinite regress, where new genes and molecular profiles seem to spring up each day, each one promising to be the "key" descriptor for the transition between health and disease. As useful as such snapshots of disease states might be in terms of diagnosis and patient substratification, what we would actually like to know is which sets of factors contribute mechanistically to, and therefore define, a clinical disease space? Moreover, knowing that a disease often only manifests in some percentage of even a high-risk population, can we account causally for the configurations of pathophysiological factors that generates clinical and epidemiological incidences of a particular disease? This last task is of critical importance when trying to make the bridge between preclinical basic science and the clinical conditions. We have already alluded to the fact that the biological proxy models that constitute the overwhelming majority of preclinical research are highly engineered systems, manipulated to meet the practical and logistical constraints of performing experimental research. One overriding factor in this time-consuming and expensive endeavor is making sure that the biological proxy model gives a strong enough "signal" for a defined experiment, i.e., it needs to generate the targeted disease process frequently enough so that a lab would not bankrupted by having "normal" outcomes in the putative disease-replicating arm. Therefore, it is not uncommon to see biological proxy models with disease penetrance of 80% or 90%, perhaps even 100%. A quick reality check must lead one to ask: what is actually being modeled here? Are the sequences of cellular and molecular events in such engineered biological systems even remotely close to what is happening in the clinical setting? True, such experimental systems are the means to identifying specific mechanistic relationships between the various moving biological parts, but the context in which they are manifest frequently bears little resemblance to the clinical scenarios that they purport to replicate. In the critical illness field, the significance of this point was brought home in a landmark study that showed that in preclinical animal models of sepsis in which control mortality was artificially high, many anti-inflammatory therapies improved mortality. However, in animal models in which the mortality in the control group was closer to that observed clinically (25–35%), not a single anti-inflammatory approach worked [50].

A dramatic example (though by no means a unique one) of this phenomenon in the context of modeling is seen in the study of necrotizing enterocolitis (NEC). We have already been briefly introduced to this disease in Chapter 4.3; we will discuss its study and putative pathophysiology in a bit more depth here. NEC is a complex, multifactorial disease that is the leading cause of gastrointestinal morbidity and mortality in the premature infant population. Despite a multitude of investigative targets, it is clear that NEC mandatorily involves prematurity, enteral feeding and a bacterial component resulting in bowel inflammation and necrosis. The pediatric research community has found it extremely challenging to create laboratory models that can reproduce comprehensively the range of pathogenic components associated with NEC, mainly related to the extreme degree of experimental perturbations required to generate the NEC phenotype *in vivo*. Examples of such maneuvers include the use of refrigeration cold-shock, hypoxic chambers, or specific, rarely clinically encountered bacteria, in order to force the development of bowel necrosis. While these measures have proven invaluable in elucidating the mechanistic factors involved in intestinal inflammation, it is clearly obvious that we do not place premature infants in refrigerators, or withhold oxygen, or intentionally force ingestion of pathogenic bacteria. Moreover, even given its clinical significance, NEC is relatively rare, affecting only about 3 out of 1000 premature births. Therefore, the gulf between the biological proxies used to study NEC and its clinical manifestation is clearly wide.

We sought to cross this gulf using the principles of dynamic knowledge representation, and by going back to the first principles of the necessary preconditions associated with NEC. We focused on formulating a *minimally sufficient unifying hypothesis of NEC* that integrated prematurity, enteral feeding and a generic bacterial component through the iterative development of an ABM that instantiated that hypothesis. As a result, we posit that the fundamental deficit in infants susceptible to NEC is immaturity of the ability of the neonatal gut epithelial cells (GECs) to manage reactive oxygen species (ROS), including those produced as a by-product of cellular respiration [51]. When this basic feature was instantiated in the NEC ABM, and then overlaid with the other recognized contributing factors, a recognizable pattern of cascading systems failure was demonstrated to be necessary for the generation of the NEC phenotype. Specifically, immature neonatal GECs had increased fragility to inflammation propagating challenges, such as metabolic stress (from feeding), decreased mucus barrier integrity and bacterial contacts. Moreover we were able to do this in a way that was able to reproduce the reported incidence of NEC, thereby providing a clinically relevant contextual picture of how these various factors interact in the production of the disease.

Having unified the known pathophysiological factors contributing to NEC, we now further refined our characterization of the disease space of NEC by adding the roles of a variable microbial component. While no single microbial species has been invoked in the pathogenesis of NEC, it is clear that many different types of bacteria can generate similar virulence phenotypes: in this fashion there can be functional groupings of bacteria that cross species. It is therefore reasonable to examine whether particular groupings of virulence capability, such as the producing of toxins or the ability to degrade intestinal mucus, might further characterize the at-risk state space of NEC. We performed just this type of analysis, describing a multidimensional state space consisting of enterocyte immaturity and various types of bacterial virulence factors that represented a continuum of patients at risk for NEC. The validity of this description is reinforced by the increasing recognition that factors involved in providing increased stability of the microbial population (i.e., producing less stimuli to produce virulent transformation) such as feeding with maternal breast milk (a "therapy" that we also modeled using ODEs [52]), appear to reduce the incidence of NEC. This process would not have been possible without the use of the NEC ABM as a means of tangible dynamic knowledge representation that could be progressively refined through the iterative addition of mechanisms. We hope that the compendium of mechanistic computational models of NEC, including the one described above and those described in Chapter 4.3, will at some point be adopted by the broader NEC research community in the search for novel diagnostic, disease-staging and therapeutic modalities [53].

We performed a similar exercise in the investigation of the pathogenesis of surgical site infections (SSIs) using an ABM, the Muscle Wound ABM (MWABM) [54]. SSIs are a significant source of postoperative morbidity and additional health-care costs [55,56], despite clinical improvements in perioperative antisepsis, antibiotics and surgical technique, and significant advances in the understanding of the molecular and cellular biology of wound healing and the impairment thereof. As with our initial identification of the mandatory components in the pathogenesis of NEC, we similarly identified that the pathogenesis of SSIs requires the following processes: tissue trauma at the site of the surgical incision, concurrent inoculation bacterial contamination, and conditions leading to a bacterial growth rate such that successful healing is impaired and the bacteria cannot be cleared leading to an abscess [55]. Also present in the current concept of the pathogenesis of SSIs is the role of excessive inflammation, which leads to a disorder of the healing processes and facilitating in the formation of an abscess. Expanding upon the mandatory conditions through the inclusion of the role of disordered inflammation, we posit the following conditions:

- That there is some relationship between the degree of bacterial contamination and the likelihood of developing an SSI, despite the baseline host characteristics. This assumption is plausible because there is no single group of patients that either never or always develops an SSI, and it is reasonable to assume that a greater level of contamination leads to a higher chance of SSI. This goes back to our being able to reproduce the clinical incidence of a disease, along with intragroup variability of outcome.
- The threshold effect of bacterial contamination is not only a function of quantity of bacteria but also of quality, i.e., specific characteristics of the type of bacteria present. Not only are certain species of bacteria known to be more aggressive in terms of the development of SSIs, many bacterial species, such as *S. aureus* and *Enterobacteria* species, among the most common bacteria seen is SSIs, have the potential to manifest a dynamically increased range of virulence factors that may predispose to SSI.
- The manifestation of bacterial virulence is a highly dynamic process, both in terms of individual microbes (i.e., through gene regulation as seen in quorum sensing) and in terms of population effects (i.e., through mutations or horizontal gene transfer and selection). This implies that the actual functional effect of a particular degree of contamination cannot be defined based on a single sample or snapshot in time, but rather can arise and progress dynamically over time. Therefore, characterizing bacterial virulence as related to the pathogenesis of SSI requires the means to capture these dynamics.

Given these assumptions, we describe the clinical state space SSI as a dynamic triangulation among the host state, the quantity of bacterial contamination and the quality in terms of virulence potential and progression of that bacterial contamination. As with our investigation into NEC, the transition from successful healing to an SSI is not a single inflection point, but rather a transition zone, or *state space*, of progressive risk for the development of an SSI. This investigation was carried out with the MWABM consisting of agents representing muscle cells, myoblasts, and various species of immune cells interacting within a simulated matrix of healthy muscle cells containing abstracted blood vessels. SSIs were generated by adding simulated bacteria to the base MWABM to identify multivariable threshold zones that mark the phase transition between healing and nonhealing/abscess formation, with the specific emphasis on characterizing the difference in thresholds between aseptic healing, in the presence of avirulent bacteria, and in the presence of bacteria with virulence potential. The key point in these simulated experiments was the demonstration that it is the dynamic expression of virulence in the bacterial population that plays a key role in increasing the likelihood of an SSI. This is a significant finding, since it suggests that it is no longer sufficient in the clinical setting to provide a general identification of a particular bacterial species (or even a single property such as antibiotic resistance) contaminating the wound. Rather, what is required is a characterization of what the microbe *might be capable of doing in the future*, i.e., its virulence potential. These features would be the target of genomic characterization of potential microbes causing SSIs, focused on identifying the existence of sequences known to be associated with virulence factors, as well as the ability to utilize horizontal gene transfer mechanisms in order to swap virulence factors [57]. In this circumstance, the circle of causality in the mechanistic pathogenesis of SSIs can be potentially closed as host activates microbe, microbe attacks host, triggering greater response to microbe [57]. Therefore, the clinical applicability of these studies on SSIs using the MWABM is in its ability to suggest dynamic pathophysiologic interactions and targeting the predisposing markers extractable from clinical samples.

INTEGRATION AND UNIFICATION: LINKING DISEASE PROCESSES, BIOLOGICAL KNOWLEDGE AND CLINICAL PHENOTYPES

We have previously discussed the potential power of developing theories of biology, theories that can not only explain behavior but also constrain what might be possible. We have also noted that this legacy of "Physics Envy" has had both good and bad effects: bad insomuch it can lead to the idea that physics must necessarily be invoked in order to comprehensively understand biology, and even that such comprehensive understanding should be the goal of bioscience; good in the sense that it provides at least something to strive for in terms of identifying fundamental driving principles for biological systems. As is consistent with the practice of Translational Systems Biology, we add the further modifier of a practical, applicable goal that can help guide us to develop useful abstractions of sufficient power. This is our concept of using dynamic computational models as means of binding together the heterogeneous landscape of observed biological phenomena. We have seen in our prior examples of the use of ABMs an emphasis on being able to unify disparate areas of biological study; we expand this concept further to link across disease processes, using the ubiquitous role of inflammation as the biological glue for our exercises in dynamic knowledge representation. Here, we provide three examples consisting of the link between inflammation and cancer, the unifying role of inflammation in intestinal disease and the continuum between inflammation and wound healing.

Bringing Together Two Biggies: Inflammation and Cancer

There is an increasing awareness of a fundamental link between inflammation and cancer, with compelling epidemiological and mechanistic evidence to support this association [58–62]. Infectious diseases that lead to chronic, nonresolving inflammation, such as Hepatitis B and C, and Human Papilloma Virus, are known to promote the development of cancer [58–60]. Conditions associated with a chronic and recurring inflammation, such as ulcerative colitis and primary sclerosing cholangitis, are well known to predispose to cancer [58–60]. Anti-inflammatory drugs, such as aspirin, have also been demonstrated to reduce overall cancer incidence [63,64].

However, when we look at mechanistic knowledge at the intersection between cancer and inflammation, the picture is not so clear [58–62]. For instance, inflammation plays a negative role in promoting the generation of genetic instability that can lead to cancer [65], but also has a positive, protective role in being able to kill early tumor cells and defend against invasion of the developed tumor [66]. We have seen this Janus-faced view of inflammation before: it is a fundamental property of the intersection between inflammation and disease [67,68]. Therefore, it is natural that we would apply Translational Systems Biology, and specifically agent-based modeling, to the study of the intersection between inflammation and cancer [69].

We focus our investigation on the most basic of questions: where do cancers come from? At first glance, this topic would seem to be far afield from the clinical emphasis we have used to characterize Translational Systems Biology. After all, by investigating the fundamental aspects of cancer, it would appear that we have taken a turn down the reductionist rabbit hole. However, we think that using the tools of Translational Systems Biology to investigate the origins of cancer, with a particular focus on inflammation, is of tremendous clinical import. To find evidence justifying our position one need only look at the history of cancer treatment over the past quarter century. If there was ever a field that manifested the Translational Dilemma it is the study of cancer, so much so that the problems inherent in the Translational Dilemma are essentially engraved in oncology's DNA to a point where they are almost assumed to be part of the nature of the discipline. For a field in which, for all intents and purposes, stage for stage mortality for solid tumors has remained constant throughout the "omics" era, in which the criteria for a "successful" drug is the extension of life measured in weeks, and where the cutting edge of clinical cancer care is an increased recognition of the benefits of palliative care over aggressive "treatment"... Well, what can one say? Clearly there seems to be some fundamental issue with how cancer is viewed and studied.

Therefore, we sought to provide a potential alternative pathway to looking at cancer in order to define novel clinical targets. Consistent with our overall approach to examining disease, oncogenesis represents that transition from health to disease that needs to be understood before trying to obtain detailed understanding of cancer biology: it is the foundation upon which all that knowledge should rest. This is the goal of Translational Systems Biology, to provide an operational strategy, grounded in fundamental principles, which can provide for a rational and methodical approach toward engineering potential therapeutics. For cancer, this means understanding where cancer comes from, so we can contextualize appropriately the highly detailed information currently being generated and recognize the fundamental forces that direct the course of the disease and how it responds dynamically to attempted interventions.

Given our penchant for definitions, we first turn to a generally accepted list of the hallmarks of cancer from Hanahan and Weinberg. There are six hallmarks: (1) sustaining proliferative signaling, (2) evading growth suppression, (3) resisting cell death, (4) enabling replicative immortality, (5) activating invasion and metastases, and (6) inducing angiogenesis [62,70]. These factors are further grouped into those concerning intrinsic properties of cancer cells, resulting from a fundamental change in their internal programming (Hallmarks 1–4) and those related to a macrophenomenon associated with a population of cancer cells, i.e., the tumor (Hallmarks 5 and 6) [58]. We further refine these categorizations by emphasizing the functional relationships and dependencies to identify fundamental driving principles in oncogenesis. This creates a hierarchy of principles, extending from the most basic aspects of biological function to higher order processes:

- *First-order process*: Promotion of genetic instability/plasticity. This process is defined by the genetic damage (DNA base pair alterations) that accumulates for each individual cell when induced genetic damage is greater than the cell's ability to repair that damage. This damage becomes generationally relevant when a cell that has accumulated damage divides. Note that this process represents changes in the DNA sequence (genotype), and not just the regulation of the gene expression network (phenotype). Therefore, alterations due to gene instability/plasticity represent a more fundamental disturbance to the function of a gene than epigenetic or signaling/regulatory.
- *Second-order process*: Functional deficits manifest at the individual cell level. These functional properties reflected in the behavior of individual cells fall into the general category of Hallmarks 1–4: promoting proliferation (either stimulating proliferation or loss of proliferation suppression), loss of mortality (dysfunction of telomerase, impairment of apoptosis), impaired damage repair (leading to increased genotypic plasticity), and loss of migration inhibition (leading to failure of multicellular tissue ordering/structure and acquisition of invasiveness, as seen resulting from epithelial–mesenchymal transition). These are all functional consequences of the genetic disturbances happening at the first-order level, and constitute the loss of evolutionarily generated control structures required to maintain the integrity of multicellular organisms. The loss of these control functions represents a shift of active and relevant evolutionary fitness/selection from the entire organism to a suborganismal level.
- *Third-order process*: Multicellular effects evident in the behavior of the tumors as a population of cells. These properties generally correlate to Hallmarks 5 and 6, and include: promoting angiogenesis, interactions with the stromal microenvironment, immune evasion, and release of potentially metastatic cells. Signaling events between tumor cells and surrounding normal tissue primarily drive these processes. Because they represent feedback between the tumor and normal tissue, many of these interactions represent hijacking of "normal" processes present in multicellular organisms, i.e., angiogenesis, tissue healing, prevention of anti-self-immune responses.

The significance of this categorization structure is that lower order processes drive and generate the higher order processes. For instance, second-order functional abnormalities result from first-order disturbances that disrupt genetic control structures; third-order processes result from the intersection between disordered cells manifesting second-order abnormalities. Therefore, focusing initial characterization of the role of inflammation in oncogenesis on the generation of genetic instability provides a fundamental grounding for the subsequent addition of more specific detail.

We identify inflammation, since it induces DNA damage, as a fundamental mechanism that drives first-order oncogenic processes. We posit that being able to understand the myriad effects of inflammation on cancer requires addressing these most fundamental aspects of the intersection between inflammation and cancer, and we utilize the principles of dynamic knowledge representation with an ABM to investigate these aspects [4,44]. Applying this approach to the question of cancer leads to a recognition of the benefits of the generative ordering schema of the hallmarks of cancer we described above. This schema provides a logical, systematic, and progressive framework that will allow subsequent layering of detail through iterative refinement. We think this is a necessary step, since given the range of pathways, genes, components and factors involved, this methodical strategy is the only way to organize the accumulating mechanistic molecular knowledge concerning inflammation and cancer.

To do this, we have developed an ABM that integrates two previously validated ABMs, one concerning oncogenesis [71] and the other inflammation [11,23,24], into a single model, the inflammation and cancer agent-based model (ICABM). The ICABM unifies the functional aspects of inflammation and oncogenesis and demonstrates that persistent, nonresolving inflammation contributes to the development of increasingly disordered epithelial cells, leading to malignant transformation. This is yet another example of the modular properties of ABMs, where the oncogenesis module consists of an epithelial cell population generated by stem cells and maintained in a dynamically stable population, a model architecture previously developed to study the process of oncogenesis in breast tissue [71]. This module incorporates several of the previously noted hallmarks of cancer: increased DNA damage or decreased DNA repair, leading to transferrable mutations in DNA repair, inappropriate migration, loss of limitation of replications, and dysregulated proliferation. The inflammation module of the ICABM is based on our old friend, the IIR ABM [11,24]. While in its original use, the IIR ABM abstracted the entire body as a single endothelial-blood interface, we now utilize its basic structure to provide a generic inflammatory milieu onto which the oncogenesis module can be placed. Given that there are multiple possible intersection points between inflammation and oncogenesis, we [4,10,16,72,73] focus on the role of generated ROS as the means of producing DNA damage arising from inflammation [58,59,61]. The ICABM uses the generation and effect of ROS to integrate the inflammation module with the oncogenesis module; this allows us to see how the forces involved in generating tumors operate within a biologically relevant environment.

Simulations with the ICABM demonstrate and confirm that inflammation provides a functional environmental context that drives mutational events in the at-risk cell population. Furthermore, these patterns of mutations lead to a cell population shift to suborganismal evolution (i.e., where tumor cells no longer recognize the initial programming from their genome, which has evolved predicated upon those cell's place within a multicellular organism), where increasingly inflammatory environments led to increasingly damaged genomes in microtumors (tumors below clinical detection size) and cancers. These simulation results lead to a conceptual framework that reconciles the dual aspects that generally separate discussion of the intersection between inflammation and cancer: (i) the generation of genetic instability at the point of oncogenesis [65] and (ii) the avoidance of immune surveillance accomplished by developing tumors [66]. The proposed relationship between these two effects is the recognition that, at least initially, developing tumors bear resemblance to their tissue of origin; therefore, they are not readily identified as nonself. However, with increasing disorder of the tumor cells, they eventually reach a point where they become recognized as a nonself invader to be responded to [58]. The same inflammatory milieu that leads to the genetic instability of the initial tumor will similarly affect the subsequent mutability (i.e., genetic plasticity as an adaptive strategy) of that tumor through its evolutionary life. The additional implication of this is that any potentially effective antitumor therapy that doesn't address reducing the inflammatory milieu is only a temporizing intervention, and highly dependent upon the degree of disorder present with the primary tumor arises (consistent with experience of biological determinism in cancer prognosis). We assert that what is needed to address this dilemma is an investigatory platform that can provide a means of representing the appropriate dynamic evolutionary context for these oncogenic processes. Furthermore, the simulation experiments demonstrate that the oncogenic environment is promoted by nonresolving inflammation, itself a dynamically stable state generated by the layered feedback architecture characterizing inflammation. We assert that only by representing the mutagenic environment (first order) in which a developed cancer will subsequently inhabit, can the higher order processes of functional cellular response (second order) and macro-population tumor effects (third order) be appropriately contextualized, evaluated, and potentially targeted for intervention. This is particularly critical in the transition from the study of specific cellular and molecular mechanisms governing the behavior of cancer with the clinical and epidemiological data that defines cancer's personal and public impact. Given the breadth of data and knowledge available at all these different levels, dynamic knowledge representation of the type provided by the ICABM can serve as a unifying platform in which the instantiation of mechanistic knowledge can be matched to, and evaluated against, population-level data. We should recognize by now that this is a fundamental aspect of Translational Systems Biology [44]. Consistent with our utilization of the power of abstraction, the ICABM is not intended to produce a comprehensive, detailed mechanistic simulacrum of a particular cancer type, in fact doing so would render it unable to make powerful, generalizable statements about the origins of cancer. The ICABM represents an evolutionary step in attempting to add the capability to add mechanistic detail to the component of such population disease/treatment models. This is particularly notable in the future study of what we have defined as "microtumors:" populations of invasive cancer cells smaller than the clinical detection threshold. These tumors represent "oncological dark matter," as by definition while they must exist they cannot be seen, and therefore cannot be studied clinically. The very early biology and natural history of cancers is highly clinically significant, with numerous implications related to balancing the benefits of surveillance and early detection with the risks and consequences of overtreatment. This phase of cancer biology can only be studied at the preclinical, basic science level, where mechanisms of spontaneous regression, immune suppression, and microenvironmental factors affecting third-order processes can be examined at a detailed cellular–molecular level. We assert that models with the capacity to dynamically represent these mechanisms while being able to generate contextualizing epidemiological data, such as the ICABM, are necessary in order to translate preclinical knowledge into actionable clinical practice.

In a similar vein but using somewhat different abstractions, we have examined the process of liver cancer (hepatocellular carcinoma) arising from chronic liver inflammation using an ABM of the liver in which inflammation leads to liver fibrosis/cirrhosis [74] and, over the long term, could lead to the formation of liver cancer (hepatocellular carcinoma; unpublished). We first created a tissue-realistic model of a two-dimensional section of the liver, the Liver Fibrosis ABM (LFABM), which includes characteristic lobule structures, hepatic portal veins as a means for cells and molecules to circulate, and a virtual "superstructure" that allows for the stiffness of this tissue to change (thereby replicating changes in tissue stiffness that occur as a function of fibrosis) [74]. This ABM could reproduce, in a tissue-realistic fashion, the inflammatory and fibrotic responses of the liver in response to noninfectious stimuli. The time-dependent patterns of inflammation and fibrosis generated by the LFABM matched those observed in rats subjected to chronic hepatotoxic chemical injury, and predicted defined changes in tissue stiffness. The LFABM was also used to carry out *in silico* clinical studies of potential anti-inflammatory and anti-fibrotic therapies, and predicted differential effects thereof [74]. An unpublished variant of this model was used to simulate hepatitis C infection. In this ABM, hepatitis C infection led, in a probabilistic fashion, to either self-resolving or chronic infections, with time courses that matched qualitatively against published time courses of the persistence of viral antigens in the circulations. In the case of the chronic infections, a portion led to the conversion of quiescent cancer stem cells into tumors, over a much longer period of time. Of these nascent tumors, a small proportion went on to be attacked by the immune system, leading to a necrotic, hypoxic core. In the setting of hypoxia, a portion of these tumors went on to form angiogenic foci, which were in essence an abstraction of the possibility of forming tumor vasculature and, hence, metastasis. These liver inflammation ABMs again highlight the multiscale nature of the inflammation ABMs. In this context, it is important to note that "multiscale" has multiple meanings. First, multiscale refers to the fact that these ABMs include molecules, cells that produce these molecules, structures/tissues that contain these cells, and, in the particular case of our LFABM, an abstracted "superstructure" that allows all of the above mechanisms to affect a macro property of overall tissue stiffness [74]. The term "multiscale model" is often also used to mean that processes encompassed within a computational model occur at very different time scales. In this regard, our liver ABM allows for rapid processes (e.g., infection with hepatitis C virus leading to inflammation within hours/days, and clearance of the virus within days/weeks), medium-duration processes (e.g., fibrosis/cirrhosis, which can take months to years), and long-term processes such as cancer, which can take years/decades. Our liver ABMs are multiscale based on all of these definitions.

A Virtual Biological Proxy Model: The Spatially Explicit General-Purpose Model of Enteric Tissue (SEGMEnT)

One of the basic principles of science, as we have noted repeatedly, is the desire to find commonalities in function and structure that underlie multiple different phenomena. This is manifest in Translational Systems Biology by the use of dynamic computational models to try to represent some fundamental level of dynamics that can bind together different clinical disease states. We have seen examples of this concerning the characterization of particular patient populations (the IIR ABM), in terms of integrating the pathophysiology of specific diseases (NEC, SSIs), as well as bridging between more general biological processes (cancer and inflammation). Now we apply that principle to the creation of a virtual version of a biological proxy model, in order to bind together different diseases occurring within a specific organ that can be represented with a high degree of spatial and structural fidelity. This last capability allows us to potentially leverage both the traditional role of histology in defining clinical disease, as well as advances in tissue and organ imaging, by connecting their observable metrics with mechanism-based cellular and molecular biology. The example we provide here involves the structure, function, and pathology of the intestinal tract [75].

As a target for this application of Translational Systems Biology, the clinical relevance of the structure–function relationship of the gut mucosa is clear: histological characterization of intestinal tissue is a mainstay in the diagnosis of intestinal disease [76]. Because there must be some fundamental biology generating these tissue phenotypes, we believe that a broad range of intestinal diseases can be unified by a view of the mucosal tissue architecture, determined by morphogenesis pathways, as being subject to a series of control modules that maintain that architecture in health, and lead to alterations seen in disease. Of these control modules, we identify inflammation as among the most clinically significant. Therefore, while the specific manifestation of gut inflammation may be different in different diseases, the same general set of processes are involved across this spectrum of diseases, and act to generate specific histological changes of the gut mucosa. With this in mind, we have developed an ABM to simulate the cellular and molecular interactions that maintain and modify the enteric mucosal architecture, the Spatially Explicit General-purpose Model of Enteric Tissue (SEGMEnT) [75]. SEGMEnT models the spatial dynamics of the crypt-villus tissue architecture maintained by the behavior of GECs as they undergo replication, migration, and differentiation, with the novel incorporation of the effect of inflammation on those morphogenic processes. SEGMEnT has a mutable morphologic 3-dimensional topology, specifically via alterations in the crypt-villus configuration and tissue architectural features (e.g., the location of the crypt-villus junction, relative and absolute sizes of the crypt and villus), allowing it to represent a range of different tissue histologies. The primary cell types represented in SEGMEnT are GECs, including their subtypes of stem cells, differentiating and mature enterocytes, and two main lineages of inflammatory cells, neutrophils and macrophages/monocytes. SEGMEnT integrates multiple functional modules including an intracellular morphogenic signaling pathway, an intracellular inflammatory signaling pathway, cell state transitions for proliferation, differentiation and movement, and spatial diffusion of morphogens to define GEC behavior for the homeostatic maintenance of the spatial architecture of the enteric mucosa.

SEGMEnT dynamically represents and integrates existing knowledge concerning homeostasis and inflammation in the intestine and provides a computational platform that can be used as an adjunct to the exploration of the cellular/molecular processes involved in intestinal wound repair, ischemia/reperfusion injury, and colonic metaplasia. In terms of validation, SEGMEnT successfully replicates the dynamically stable morphology and cellular populations of the healthy ileum, while qualitatively producing realistic spatial distributions of molecular signaling gradients using existing qualitative histological criteria [77–84]. SEGMEnT also reproduces the effects of inhibition of various signaling pathways; i.e., simulating the behaviors of morphogen knockout and silencing experiments [77–84]. Given our modeling paradigm of being able to represent a robust healthy state, SEGMEnT is able to withstand certain acute perturbations (local tissue injury, ischemia/reperfusion) as would be expected from normal intestinal tissue. In terms of long-term behavior, SEGMEnT can also reproduce a specific chronic pathological state associated with ongoing, low-level inflammatory stimulation (colonic metaplasia). Furthermore, simulations reinforce the protective role of enterocyte sloughing in enteric ischemia-reperfusion and suggest a novel cross-talk nexus between inflammation and epithelial patterning in the continuum effect of inflammation on the genesis of colonic metaplasia.

We see manifest in SEGMEnT many of the features seen in a biological proxy model: it is able to generate a range of behaviors, produces output that is clinically relevant in terms of tissue structure, and is able to replicate organ-scale molecularly detailed representations (the high-performance computing version of SEGMEnT can simulate up to a meter of intestinal tissue at the molecular level), but with all the advantages of an *in silico* model providing dynamic knowledge representation. As such, it demonstrates the power of the underlying conceptual model, by being able to bind together a heterogeneous set of disease processes, and do so in a context that can account (at least for now) for the cross talk between those included pathways. Calling back to the modular nature of ABMs, so too does SEGMEnT's modular control structure offer rational directed paths forward, by providing a common end effector for investigations into the role of the gut microbiome, or the generation of colon cancer, or the effects of diet on modulating metabolic diseases. SEGMEnT's level of detail has come a long way from the abstractions of the initial IIR ABMs, but its detail purposefully added, with a specific clinical phenotype being targeted, and not just for the sake of adding more mechanisms. In this fashion, SEGMEnT well represents the tenets of Translational Systems Biology as a means of providing a systematic strategy for sequentially adding increased mechanistic knowledge when appropriately called for.

Capturing the Continuum from Injury to Repair: *In Silico* Trials of Wound Healing

We started our survey of the use of agent-based modeling in Translational Systems Biology by describing our initial simulations directed at conducting *in silico* clinical trials. As we noted, this is consistent with the emphasis of Translational Systems Biology on examining clinically relevant situations. We now circle back to the performance of *in silico* clinical trials as we examine the close relationship between acute injury and wound healing. This relationship should be obvious: injuries need to heal, and therefore it would make sense that the cellular and molecular processes involved in the acute response to injury would be directly responsible for initiating the healing process. However, when we embarked on these investigations, we were surprised that the mechanistic chain of cause and effect leading from injury to healing was not more appreciated. We posit that this is yet another example of the compartmentalized nature of biomedical research, with injury and healing treated as separate processes by their respectively different investigatory domains: shock and trauma for injury, chronic wound care for healing. We also believe it was our employment of mechanistic modeling, and the intrinsic need to note cause and effect formally and explicitly, that made the continuum between injury and healing immediately obvious to us. It was quickly evident that many cellular and molecular mediators are shared between acute inflammation and healing; for instance, the anti-inflammatory mediators that limit and contain the propagation of the pro-inflammatory response, such as interleukin-10 and transforming growth factor-β1(TGF-β1) are themselves growth factors, with TGF-β1 especially being considered a key (perhaps *the* key) mediator in wound healing [85]. Wound healing is also an intrinsically spatial process, as damaged tissue is removed and replaced by surrounding "normal" tissue. Therefore, ABMs of wound healing represented a natural direction of development arising from the early inflammatory ABMs. Wound-healing ABMs have been used to shed basic insights on the spatial nature of skin wounds and their healing [34,57], to represent the mechanistic pathophysiology of diabetic wounds and to posit potential mechanistic targets for therapeutics development [33], offer the potential for personalized medicine by modeling individual responses to injury and therapy in vocal fold trauma [39,86], and simulate the pathophysiology and potential treatment of pressure-related decubitus ulcers in patients with spinal cord injury (SCI) [87].

The diabetic wound ABM [33] was used to determine the phenotypic effects of under-activation of latent TGF-β1 and overproduction of tumor necrosis factor-α (TNF-α), both associated with diabetes. This ABM generated a host of system-level features characteristic of diabetic ulcers. Moreover, this ABM was used to test *in silico* the effects of both current therapies for diabetic ulcers (namely wound debridement and treatment with platelet-derived growth factor) and novel interventions (e.g., inhibition of TNF-α or addition of TGF-β1) [33]. This ABM embodied one of the core principles of Translational Systems Biology that we have mentioned previously, namely that a simulation has to be able to capture the behavior of a healthy system before attempting to model what happens in a disease setting. Though we refer to this model as a "diabetic wound ABM," this is something of a misnomer: under baseline conditions, this model could recapitulate the healing trajectories of normal skin. Only when the model was perturbed in specific ways—namely tuning up TNF-α production or tuning down TGF-β1 activation—did it exhibit features characteristic of diabetic wounds. Thus, the model could be extended in the future to cover other aspects of skin injury, for example, burns or infected wounds, starting with the baseline configuration. Another possibility would be to account for changes in wound healing as a function of age or other disease process separate from diabetic wounds. On the other hand, the ability of the ABM to capture behaviors consistent with diabetic wounds based on single, explicit changes may help us gain an understanding of the derangements that underlie chronic, nonhealing diabetic wounds.

The diabetic wound ABM was our first attempt at integrating key inflammatory mechanisms with key, well-vetted mechanisms of wound healing in order to test the hypothesis that such a structure would lead to viable predictions with regard to a clinically realistic problem. As we describe above, another goal of that study was generation of simulated cohorts of patients, similar to those we created in the sepsis ABM described above, as well as in the context of our *in silico* clinical trials using ODE models (detailed in Chapter 4.3). We next sought to examine the complete opposite end of the spectrum, by creating individual-specific, mechanistic models of inflammation and wound healing. This was one of the mainstay concepts of Translational Systems Biology, and so achieving viable predictions and insights from such a model would represent a major milestone with regard to clinically useful modeling. However, we were faced with a major challenge fairly immediately: ideally, we would structure a model that recapitulates inflammation in an individual in response to a given stimulus, but how does one go about creating a prospective inflammatory stress in a person? More specifically, if we wish to study the response to injury, we cannot, ethically, injure a person and observe (and hopefully predict) the ensuing time course of inflammation. This conundrum led us to one of our many interdisciplinary collaborations, this time with a group that was studying the trauma that the vocal folds experience when people speak too loudly and for too long (phonotrauma) [88]. We hypothesized that phonotrauma to the vocal folds was a reasonable representation of trauma to any other tissue. Our collaborators were able to collect secretions produced near the vocal folds, and we made the further approximation that these secretions represented the inflammatory response produced in the vocal fold tissue itself, much like our approximation that inflammatory mediators in the blood represent the underlying, multi-tissue/organ inflammation in the setting of critical illness. Based on these approximations, we created an ABM of vocal fold injury and inflammation, which we used to create personalized sets of models by calibrating parameters using data on inflammatory mediator levels in laryngeal secretions of individual human volunteers subjected to experimental phonotrauma [89]. Individual-specific computational simulations were created based on baseline levels of these mediators as well as at 1 and 4 h after phonotrauma. These simulations generally predicted the levels of inflammatory mediators at much later time points (24 h), and were used as the basis for simulated therapy [89]. This was done in a way that was different from our prior approaches. Specifically, we looked at individual human subjects that, after generating vocal fold trauma by very loud phonation were randomized into one of three possible (nonpharmacologic) treatments—normal speech (the worst option for injured vocal folds), voice rest (the "standard" prescription for injured vocal folds), or resonant voice exercise (an experimental concept introduced by our collaborators) [88]. We used our ABM to show that we could predict quantitatively the inflammatory outcomes at 24 h, and predict qualitatively the "damage" outcomes at the same time point. Perhaps the most important outcome of this study is that we predicted what would happen to individuals if they had been randomized to a different group from the one into which they had actually been randomized. The reason this was important is that, for the first time, the possibility was raised of determining, prospectively, the optimal treatment for a person based on limited data in combination with a model. This approach was later recapitulated in our data-driven modeling studies of human trauma (see Chapter 4.2).

Based on this growing experience in modeling inflammation and wound healing, as well as in modeling the responses of individual patients, we sought to expand this approach in several ways. First, we sought to address another important clinical problem, and so we focused on pressure ulcers (PUs). PUs affect 2.5 million US acute care patients and cost up to $1 billion per year [90]. They are a significant source of morbidity in both hospitalized patients and community-dwelling individuals with impaired mobility, and are especially common in individuals with spinal cord injury (SCI) [91]. Given that SCI is associated with elevation of systemic inflammatory markers [92–94], and that PU are thought to arise from pressure-induced ischemia, reperfusion injury, and/or deformation-induced cellular damage [95], we reasoned we could use dynamic modeling to gain insights into this process. Furthermore, we know that local blood flow and other key properties of patients with SCI are altered due to the lack of innervation in the affected tissues, and so another key goal of our study involved trying to link these tissue blood flow changes to the propensity to produce PU post-SCI. Finally, we had a growing number of inflammation/wound-healing ABMs, but none of them was set up to simulate the realistic appearance of lesions on the skin. Therefore, another key goal of these subsequent studies was the generation of tissue-realistic simulations. Why? Besides the inherent value in being able to validate key mechanistic assumptions at the level of the pattern that is generated, there is an important translational possibility inherent in such an effort. After all, images are data, too; images of wounds are data that can be obtained quite frequently and noninvasively, and thereby could, at some point, be used in concert with a tissue-realistic ABM to predict the outcomes of individual patients. Only recently have experimental methodologies emerged that may allow for the study of the time courses of wound healing in humans [96], but these approaches are limited in that time courses of primary samples from humans with chronic wounds are difficult to collect without disturbing the very process being measured. Also, if we pursue the approach outlined above in our vocal fold phonotrauma study [89], we might be able to predict the outcomes of PU in individual SCI patient under different treatment regimens.

Our first foray into simulating post-SCI PU involved a relatively simple ABM of inflammation and wound healing, connected to a relatively simple equation-based model of tissue blood flow and resistance thereto; this was a so-called "hybrid model," an area of very active investigation in the modeling field [87]. This hybrid model was generated because it provided the best features of a stochastic, multistate model (the ABM portion) with the best features of model that uses the mean-field approximation with regard to blood flow (the equation-based portion). This modeling study led to the following advances: the ability to blend equations and agents in a single model; the ability to model the behavior of patient cohorts (since the hybrid model predicted greater propensity to ulcerate in SCI patients versus uninjured control subjects); and the ability to create individual-specific models based solely on blood flow data that were obtained noninvasively [87].

Based on this hybrid model, we next generated tissue-realistic ABM of PU formation (the Pressure Ulcer Agent-Based Model, or PUABM; submitted for publication). The agents in the PUABM represent either cell types or cellular structures: neutrophils, macrophages, tissue cells, and blood vessels. Data layers are employed to represent mediators in a computationally efficient manner (e.g., diffusible cytokines, free radicals, oxygen, xanthine oxidase, and exogenously administered drugs). At the molecular level, the model was calibrated to *in vivo* studies of ischemia/reperfusion injury in rat epidermis [97] as well as inflammatory dynamics reported in the literature. Included explicitly is the forward-feedback loop of inflammation to damage to inflammation that has served as the core motif of our prior simulations of inflammation in both systemic and local contexts [44,98,99]. The PUABM replicates visual morphology associated with the development and resolution of post-SCI PU by simulating vascularized soft tissue overlaying a bony prominence (the clinically recognized "pressure points" at which PU typically develop) and the effects of repeated ischemia/reperfusion (representing the turning of a person with SCI in bed) on such an area of tissue. The model output matched clinical ulcers, generating irregular shapes and jagged edges, as well as distinct, secondary foci of inflammation that could progress to ulcers, despite initial conditions simulating a circular area of bony protrusion. Also recapitulating clinical outcomes, the model reaches two distinct end points when simulated from the same initial parameters, which can be interpreted as worsening inflammation/damage versus inflammation that does not resolve but does not worsen, either. We leveraged thousands of model simulations to explore the root of this phenomenon and predicted at what time an ulcer's fate is determined. We also demonstrate the utility of the PUABM as a platform for *in silico* clinical trials of strategies for prevention and therapies for treatment of PU post-SCI. This model suggests that the reason treatments thus far have been ineffective is that they have been applied too late. From a diagnostic standpoint, our simulations suggest that the most important predictors of ulcer formation are tissue oxygen levels and the levels of pro-inflammatory mediators.

Thus, wound oxygenation may be a potential therapeutic avenue for post-SCI PU. To date, studies of hyperbaric oxygen treatment have not proven successful in treating PU [100]. This failure may be explained by the timing issues mentioned above: we hypothesize that oxygenation is key to prevent ulcers from forming, but less helpful once they do.

Model parameters were calibrated to cell-level phenomena (e.g., life spans) and tissue phenotypes relevant to healthy conditions (e.g., diffusion rates were adjusted to ensure adequate oxygenation to tissue that was not experiencing pressure). It is notable then, that when pressure was applied, ulcers not only developed but did so in a comparable time frame to—and with morphology that mimicked—clinical outcomes: irregular shapes emerged despite being simulated as starting from a perfectly round ischemic area. Simulations also stochastically reproduced a range of pathological outcomes similar to that seen at the bedside: some PU incurred (or were associated with) less damage than others. The value of the PUABM lies in its ability to recapitulate a broad range of pathology from a single set of parameters, and to cover the full range of outcomes by varying only a few parameters. This model elucidates a very promising pathway to personalized medicine. Better measurements at the bedside and further analysis could allow the model be used to identify individual wound-healing phenotypes and trajectories, determine appropriate treatment course, and design and test new treatment regimens. For example, the role of temperature on ulcer formation could be implemented in the model as a parameter affecting the rates of molecular interactions. Model outcomes could then be compared to clinical studies exploring the effects of cooling cushions on ulcer progression, like those in Refs. [101,102], and also used to make predictions about therapeutic combinations.

One could imagine a model with readout like this one being used for many disease applications in which the outcome is often a visual pattern (wounds and other skin lesions (e.g., psoriasis, melanoma), histological and immunohistological biopsies, and other medical imaging modalities (e.g., magentic resonance imaging [MRI] or computer-assisted tomography [CAT] scan). The PUABM also represents an extension of our prior work on tissue-realistic modeling, e.g., in the liver [74] described above. In addition, as in our hybrid model described above, blood flow data could be integrated with clinical images to further improve diagnosis or treatment, in essence comprising a novel, multiscale diagnostic platform for post-SCI ulcer formation.

INTEGRATION AND UNIFICATION: BRINGING TOGETHER BIOMEDICAL KNOWLEDGE BY PUTTING HUMPTY DUMPTY TOGETHER AGAIN

We have seen throughout this chapter how the development of our biomedical ABMs has mirrored the developments in Translational Systems Biology, primarily serving as a means to integrate knowledge. Having progressed through patient cohorts (sepsis), to specific diseases (NEC and SSI), to integration of biological processes (oncogenesis and inflammation), to high-resolution organ models, the natural end point is the research community itself. We have at multiple points in this chapter harkened back to the modular and multiscale nature of ABMs; now we have reached a point where we can discuss how ABMs, or more correctly, the architecture of an ABM, can be used to integrate community-wide knowledge, in essence to provide a mechanism for community-wide dynamic knowledge representation.

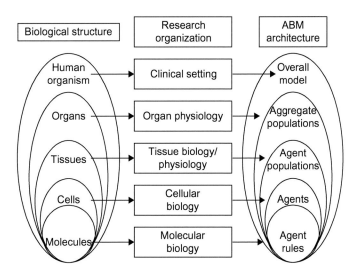

FIGURE 4.4.1 **Mapping from Biological Organization to the Research Community Structure to the structure of ABMs.** This schematic demonstrates the nested hierarchical organization of biological systems: first, how each level of biology is studied by distinct and discrete research communities with little translation across organizational levels, and second, how the knowledge generated by each research community can be reconstructed using the structure of an ABM. *Reprinted from [103].*

The kernel of this idea came relatively early: in 2006, we published a report entitled: "Concepts for Developing a Collaborative Agent-based Model of Acute Inflammation" [103]. In this paper, we noted that the structure of an ABM actually better mirrored that of a biological organism than did the disjointed biomedical research community, but that this ABM structure could be used to reconstruct the pieces broken apart by the existing reductionist research process. Figure 4.4.1 illustrates the mapping between the nested organizational hierarchies present in biological systems; the discretized, distinct, and disjointed structure of the biomedical research community; and the modular structure of an ABM, which can be used to reintegrate and synthesize the knowledge generated from that research community [103]. At that point in time, we introduced a rudimentary "grammar," that could map to biological statements about structure and function, with the intent that this grammar would be used to translate biomedical knowledge into rules for ABMs. This grammar turned out to be a "proto-ontology" for acute inflammation as well as a template for future formalized representation of the structure of an ABM; as such it led into a foray into the world of ontologies and knowledge representation.

There has been considerable interest in the bioinformatics community in providing formal structure to the knowledge generated by biomedical research. The benefits are fairly clear:

- Such a formal structure provides an organizational structure to the established corpus of knowledge.
- It facilitates navigation within a particular domain to identify points of knowledge intersection, and facilitate the advancement of knowledge within the scope of that domain.
- It provides a frame of reference into which new terminology could be integrated.
- It provides a logical structure that would be amenable to automated inference concerning relationship identification and representation.

Ontologies are knowledge classification systems that provide a structured vocabulary and taxonomy for a particular scientific domain [104]. The structure and descriptive capacity of ontologies emphasize taxonomic class structures, their properties, and the relationships between the constitutive concepts within the domain. The descriptive capacity of ontologies is currently defined by standards established in the Web Ontology Language (OWL) and the Resource Description Format (RDF). These standards consist of a collection of facts and axioms that define classes, properties, individual instances, and their interrelationships in the ontology. These are systems with the expressiveness of description logics, a decidable fragment of first-order logic with attractive computational properties, and therefore allow the use of automated inference to identify and prove relationship structures between concepts (i.e., they do not rise to the level of expressiveness that would allow them to be covered by Godel's Incompleteness Theorems).

Biomedical ontologies follow the conventions put forth in OWL and RDF. One of the most successful bioontologies is the Gene Ontology (GO) [105], which provides a classification system for the ever-increasing knowledge concerning genetic information. Bio-ontologies are currently found in the online repository BioPortal [106], which is managed and curated by the National Center for Biomedical Ontologies (NCBO) [107].

However, despite their usefulness, ontologies/bio-ontologies have a significant limitation in terms of their expressiveness, i.e., the types of statements that can be made. They serve primarily for categorization, able to express a proposition denoting a dynamic relationship between two concepts, but the conditions and circumstances in which that dynamic relationship takes place cannot be expressed. There is a need to expand expressive capacity to capture actions or functions: this requires the ability to express *rules*. This critical deficiency must be corrected given the goal of being able to generate a dynamic representation of a knowledge structure, a fact the Semantic Web research community has already recognized. They have responded to the limitations of OWL with ongoing development in languages such as Rule Markup Language (RuleML) http://ruleml.org/ and Semantic Web Rule Language (SWRL) http://www.w3.org/Submission/SWRL/. While these efforts continue, we adopted another strategy that consisted of integrating current ontology-based knowledge representation with suitable modeling methods, specifically agent-based modeling.

Knowing what we know about ABMs allows us to realize that the metastructure of an ABM essentially corresponds to an ontology. Both ABMs and ontologies share an object-oriented architecture with properties such as state-defining variables, inheritance, and child–parent relationships. However, since ABMs are dynamic models that actually need to run, they contain orthogonal object class properties that define functional capabilities, physical–structural relationships, and an operator sequence/algorithm in addition to standard ontology identity relationships. In short, ABMs contain expressions of the mechanistic *rules* currently lacking in ontologies. As noted earlier in this chapter, while the rules for simulation agents in an ABM are usually logical or algebraic statements, there is no reason that the rule set cannot be a mathematical model in itself. There are multiple examples of embedding complex mathematical models, such as an ODE or dynamic network model, within a cell-level simulation agent [5–7,35,42,43]. These examples reinforce the potential unifying role of agent-based modeling as a means of "wrapping" different simulation methodologies. Toward that end, we propose that the metastructure of an ABM can be used as a template in which structured biomedical knowledge can be concatenated to facilitate the instantiation of mechanistic hypotheses.

We developed an agent-based modeling format (ABMF) that can be used to concatenate ontological descriptions of biological components and functions to facilitate the development of an executable layer of knowledge representation [14]. Figure 4.4.2 is a schematic depicting the core module of the ABMF, essentially mapping the structure of an ABM to that of a bioontology. The ABMF utilizes terms from standard BioPortal ontologies and places them into modules that incorporate the properties needed to construct an ABM. The modular structure of the ABMF takes advantage of the multiscale representation capability associated with the structure of an ABM: the intrinsic ability to

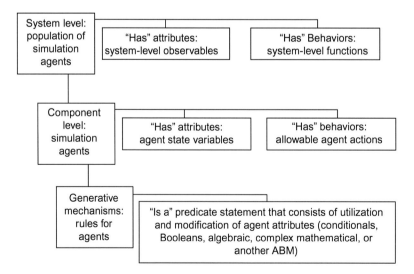

FIGURE 4.4.2   **ABMF module.** The ABMF module incorporates three hierarchical levels of representation centered on the simulation agent level. This level corresponds to the classical agent level in an ABM, in which the system-level corresponds to agent population behavior (including so-called "emergent" phenomenon), whereas the lowest level of organization, generative mechanisms, corresponds to agent rules. It should be noted that agent rules can be any formal model system, including another ABM. This property gives the ABMF a potentially recursive structure that allows nesting of ABMF modules. Also, since the generative rules can be any type of mathematical model, the ABMF is also capable of being a pathway to hybrid computational models that concurrently employ multiple modeling and simulation methods. *Reprinted from [108].*

represent at least three levels of system organization. The components of the ABMF module possess properties that are organized in a series of orthogonal hierarchies that draw upon the structures of ontologies and object-oriented programming. These hierarchies represent different class structures that can be populated with terms extracted from BioPortal ontologies.

Drawing upon the structure of an ABM, we will focus on a brief description of the properties present at each organizational level. The first of these are the *attributes* of the simulation agent. These *attributes* correspond to the state variables correlated to observable characteristics of the reference system's components being represented by the simulation agents. Syntactically, these represent *noun* and *adjective* characteristics associated with the simulation agent. These include physical components (such as proteins for cells or cells for tissues) or structural properties (such as shape and orientation). Next, recognizing that the eventual goal is the generation of a dynamic model, the simulation agent contains *behaviors*; these correspond to a list of actions or functions that the simulation agent is capable of performing. *Behaviors* correspond syntactically to *verbs*. It is now immediately evident that we are dealing with two orthogonal descriptive hierarchies/ontologies. Thus far, the *attributes* and *behaviors* are characteristics that can be associated with a simulation agent using first-order logic. However, there is a need to be able to instantiate dynamics into the simulation and this requires the formulation of a series of *rules* that determine computational actions. Here the ABMF draws upon the ABM structure such that the *rules* governing the simulation agent level represent a lower level of organization corresponding to the generative mechanisms determining the behavior of the biological reference component. The internal logic of these *rules* can be represented with a mathematical formulation that represents the lowest (=most basic) level of knowledge regress in a particular model; these formulations can be conditional or logical rules or mathematical models in of themselves. The variables in the rule formulation are drawn from the previously defined *attributes* of the simulation agent class. Alternatively, the modular structure of the ABMF is such that the rules can be another ABM and recursively parsed into the previously defined properties: *attributes* and *behaviors*.

There is not only a linkage between the simulation agent level and the generative mechanism level; this connection is common to all mathematical models and not unique to agent-based modeling. Rather, an ABM (and therefore the ABMF) has an additional, higher level of organizational representation: that of aggregated simulation agent population behavior. This is the classical output level of an ABM: those observable system properties that cannot be parsed simply into summed component actions. The *behaviors* at this level are those associated with the aggregated system's observables; its *attributes* have a similar relationship to the constitutive simulation agents, but also feedback on their interactions by forming the context for its constitutive simulation agents (i.e., the environmental topology). Therefore, there is an inescapable influence of higher-level context on the behaviors of the constitutive components, reinforcing the importance of semantics in the organizational structure associated with multihierarchical biological systems. It should also be noted that the modular structure of the ABMF allows for a recursive description of biological systems: the generative rule set for a particular level of simulation agent can itself be an ABM. It is readily evident how this structure maps onto biological systems, and allows the ABMF a high degree of expressiveness and the ability to create hybrid simulation models. This organizational recursion has been noted as a property of biological systems [109].

The representation of a conceptual model in the ABMF provides a knowledge structure that can potentially be transformed into executable models, but it should also be noted that the ABMF is not in itself a modeling platform. There is still a significant gulf between the collection of appropriate terms referring to a biological hypothesis and the ability to construct a computer simulation of that hypothesis. This programming task is nontrivial, and requires specific expertise that further complicates the efficiency of the workflow. We suggest that another computational description of an "agent" can be used to facilitate this process; we will address this concept in Chapter 4.5.

But before we leave agent-based modeling, after this long excursion, we think it would be beneficial to summarize the key points in our discussion about this method. Notably, the evolution of Translational Systems Biology has been intimately linked to the ever-increasing expansion of agent-based modeling. This is particularly true in terms of the role of dynamic knowledge representation, and its place within the Scientific Cycle in terms of providing an *in silico* means of verifying or falsifying mechanistic hypotheses. With its intuitive mapping to biological systems, agent-based modeling has played a key role in identifying how such dynamic knowledge representation could be applied to particular biomedical questions. The ability to ask these questions, with a perpetual eye on the overall clinical picture, is a central hallmark of the use of ABMs in Translational Systems Biology. Finally, our recognition that, to a great degree, ABMs map also to the structure of the research community, our studies in potentially using the metastructure of ABMs as a pathway toward semiautomated model construction, directly led to our work on developing a high-throughput dynamic computational modeling solution to the Translational Dilemma. It is to this solution that we turn in the next chapter.

RESOURCES FOR AGENT-BASED MODELING AND SUGGESTED READING

Agent-based modeling is very much a contact sport, in which the fastest way to appreciate the potential benefits of the method is by creating ABMs. There are several toolkits that have been used for agent-based modeling. There are several reasons for the development of these specialized toolkits. These reasons include the need to emulate parallel processing to represent the actions of multiple agents within populations, dealing with associated execution concurrency issues within those populations, establishing means of defining model topology (i.e., agent interaction neighborhood), and the development of task schedulers to account for the multiple iterations that constitute an ABM run. As a result of these issues, along with the case that many researchers who utilize ABMs are not trained computer scientists or programers, many biomedical ABMs are created using existing ABM development software packages. These agent-based modeling environments attempt to strike a balance between representational capacity, computational efficiency, and user-friendliness. A noncomprehensive list of such ABM toolkits includes Swarm (http://www.swarm.org/index.php/Swarm_main_page), Mason (http://cs.gmu.edu/~eclab/projects/mason/), RePast (http://repast.sourceforge.net/), NetLogo (http://ccl.northwestern.edu/netlogo/), StarLogo (http://education.mit.edu/starlogo/), and SPARK (Simple Platform for Agent-based Representation of Knowledge www.pitt.edu/~cirm/spark [110]). All these platforms represent some trade-off among the triad of goals mentioned above. For an excellent review and comparison of many of these agent-based modeling toolkits, see Ref. [111].

The following book represents an excellent introduction to agent-based modeling that will help novices learn how to make ABMs using NetLogo.

Agent-Based and Individual-Based Modeling: A Practical Introduction, by Steven F. Railsback and Volker Grimm. Our hope is that the work described and referenced herein will give the reader an appreciation both for ABMs in particular and for the translational applications of dynamic mechanistic modeling in general.

References

[1] An G. A model of TLR4 signaling and tolerance using a qualitative, particle-event-based method: introduction of spatially configured stochastic reaction chambers (SCSRC). Math Biosci 2009;217(1):43–52.

[2] Bankes SC. Agent-based modeling: a revolution? Proc Natl Acad Sci USA 2002;99(Suppl. 3):7199–200.

[3] Bonabeau E. Agent-based modeling: methods and techniques for simulating human systems. Proc Natl Acad Sci USA 2002;99(Suppl. 3):7280–7.

[4] Hunt CA, Ropella GE, Lam TN, Tang J, Kim SH, Engelberg JA, et al. At the biological modeling and simulation frontier. Pharm Res 2009;26(11):2369–400.

[5] Walker DC, Southgate J. The virtual cell—a candidate co-ordinator for 'middle-out' modeling of biological systems. Brief Bioinform 2009;10(4):450–61.

[6] Zhang L, Athale CA, Deisboeck TS. Development of a three-dimensional multiscale agent-based tumor model: simulating gene-protein interaction profiles, cell phenotypes and multicellular patterns in brain cancer. J Theor Biol 2007;244(1):96–107.

[7] Santoni D, Pedicini M, Castiglione F. Implementation of a regulatory gene network to simulate the TH1/2 differentiation in an agent-based model of hypersensitivity reactions. Bioinformatics 2008;24(11):1374–80.

[8] Fallahi-Sichani M, El-Kebir M, Marino S, Kirschner DE, Linderman JJ. Multiscale computational modeling reveals a critical role for TNF-alpha receptor 1 dynamics in tuberculosis granuloma formation. J Immunol 2011;186(6):3472–83.

[9] Zaborin A, Romanowski K, Gerdes S, Holbrook C, Lepine F, Long J, et al. Red death in *Caenorhabditis elegans* caused by *Pseudomonas aeruginosa* PAO1. Proc Natl Acad Sci USA 2009;106(15):6327–32.

[10] Hunt CA, Ropella GE, Yan L, Hung DY, Roberts MS. Physiologically based synthetic models of hepatic disposition. J Pharmacokinet Pharmacodyn 2006;33(6):737–72.

[11] An G. Introduction of an agent-based multi-scale modular architecture for dynamic knowledge representation of acute inflammation. Theor Biol Med Model 2008;5(1):11.

[12] An G, Wilensky U. From artificial life to *in silico* medicine: NetLogo as a means of translational knowledge representation in biomedical research Adamatsky A, Komosinski M, editors. Artificial life in software (2nd ed.). London: Springer-Verlag; 2009. p. 183–209.

[13] An G. Closing the scientific loop: bridging correlation and causality in the petaflop age. Sci Transl Med 2010;2(41):41ps34.

[14] Christley S, An G, editors. A proposed method for dynamic knowledge representation via agent-directed composition from biomedical and simulation ontologies: an example using gut mucus layer dynamics. Spring simulation multiconference/agent-directed simulation symposium. Boston, MA; 2011.

[15] Graner F, Glazier JA. Simulation of biological cell sorting using a two-dimensional extended Potts model. Phys Rev Lett 1992;69(13):2013–6.

[16] Engelberg JA, Ropella GE, Hunt CA. Essential operating principles for tumor spheroid growth. BMC Syst Biol 2008;2(1):110.

[17] Wendelsdorf KV, Alam M, Bassaganya-Riera J, Bisset K, Eubank S, Hontecillas R, et al. Enteric Immunity Simulator: a tool for *in silico* study of gastroenteric infections. IEEE Trans Nanobioscience 2012;11(3):273–88.

[18] Reynolds CW, editor. Flocks, herds, and schools: a distributed behavioral model in computer graphics. SIGGRAPH'87; 1987.

[19] Lipniacki T, Paszek P, Brasier AR, Luxon BA, Kimmel M. Stochastic regulation in early immune response. Biophys J 2006;90(3):725–42.

[20] Lipniacki T, Paszek P, Marciniak-Czochra A, Brasier AR, Kimmel M. Transcriptional stochasticity in gene expression. J Theor Biol 2006;238(2):348–67.

[21] Vodovotz Y, Clermont G, Hunt CA, Lefering R, Bartels J, Seydel R, et al. Evidence-based modeling of critical illness: an initial consensus from the Society for Complexity in Acute Illness. J Crit Care 2007;22(1):77–84.

[22] Grimm V, Revilla E, Berger U, Jeltsch F, Mooij W, Railsback S, et al. Pattern-oriented modeling of agent-based complex systems: lessons from ecology. Science 2005;310:987–91.

[23] An G. Agent-based computer simulation and sirs: building a bridge between basic science and clinical trials. Shock 2001;16(4):266–73.

[24] An G. *In silico* experiments of existing and hypothetical cytokine-directed clinical trials using agent-based modeling. Crit Care Med 2004;32(10):2050–60.

[25] Mansury Y, Diggory M, Deisboeck TS. Evolutionary game theory in an agent-based brain tumor model: exploring the 'Genotype-Phenotype' link. J Theor Biol 2006;238(1):146–56.

[26] Deisboeck TS, Berens ME, Kansal AR, Torquato S, Stemmer-Rachamimov AO, Chiocca EA. Pattern of self-organization in tumour systems: complex growth dynamics in a novel brain tumour spheroid model. Cell Prolif 2001;34(2):115–34.

[27] Chen S, Ganguli S, Hunt CA. An agent-based computational approach for representing aspects of *in vitro* multi-cellular tumor spheroid growth. Conf Proc IEEE Eng Med Biol Soc 2004;1:691–4.

[28] Thorne BC, Bailey AM, Benedict K, Peirce-Cottler S. Modeling blood vessel growth and leukocyte extravasation in ischemic injury: an integrated agent-based and finite element analysis approach. J Crit Care 2006;21(4):346.

[29] Tang J, Ley KF, Hunt CA. Dynamics of *in silico* leukocyte rolling, activation, and adhesion. BMC Syst Biol 2007;1:14.

[30] Tang J, Hunt CA, Mellein J, Ley K. Simulating leukocyte-venule interactions—a novel agent-oriented approach. Conf Proc IEEE Eng Med Biol Soc 2004;7:4978–81.

[31] Bailey AM, Thorne BC, Peirce SM. Multi-cell agent-based simulation of the microvasculature to study the dynamics of circulating inflammatory cell trafficking. Ann Biomed Eng 2007;35(6):916–36.

[32] Bailey AM, Lawrence MB, Shang H, Katz AJ, Peirce SM. Agent-based model of therapeutic adipose-derived stromal cell trafficking during ischemia predicts ability to roll on P-selectin. PLoS Comput Biol 2009;5(2):e1000294.

[33] Mi Q, Riviere B, Clermont G, Steed DL, Vodovotz Y. Agent-based model of inflammation and wound healing: insights into diabetic foot ulcer pathology and the role of transforming growth factor-beta1. Wound Repair Regen 2007;15(5):671–82.

[34] Walker DC, Hill G, Wood SM, Smallwood RH, Southgate J. Agent-based computational modeling of wounded epithelial cell monolayers. IEEE Trans Nanobioscience 2004;3(3):153–63.

[35] Adra S, Sun T, MacNeil S, Holcombe M, Smallwood R. Development of a three dimensional multiscale computational model of the human epidermis. PLoS One 2010;5(1):e8511.

[36] Broderick G, Ru'aini M, Chan E, Ellison MJ. A life-like virtual cell membrane using discrete automata. In Silico Biol 2005;5(2):163–78.

[37] Pogson M, Holcombe M, Smallwood R, Qwarnstrom E. Introducing spatial information into predictive NF-kappaB modelling—an agent-based approach. PLoS One 2008;3(6):e2367.

[38] Pogson M, Smallwood R, Qwarnstrom E, Holcombe M. Formal agent-based modelling of intracellular chemical interactions. Biosystems 2006;85(1):37–45.

[39] Ridgway D, Broderick G, Lopez-Campistrous A, Ru'aini M, Winter P, Hamilton M, et al. Coarse-grained molecular simulation of diffusion and reaction kinetics in a crowded virtual cytoplasm. Biophys J 2008;94(10):3748–59.

[40] Troisi A, Wong V, Ratner MA. An agent-based approach for modeling molecular self-organization. Proc Natl Acad Sci USA 2005;102(2):255–60.

[41] Dong X, Foteinou PT, Calvano SE, Lowry SF, Androulakis IP. Agent-based modeling of endotoxin-induced acute inflammatory response in human blood leukocytes. PLoS One 2010;5(2):e9249.

[42] Hoehme S, Drasdo D. A cell-based simulation software for multi-cellular systems. Bioinformatics 2010;26(20):2641–2.

[43] Christley S, Alber MS, Newman SA. Patterns of mesenchymal condensation in a multiscale, discrete stochastic model. PLoS Comput Biol 2007;3(4):e76.

[44] Vodovotz Y, Csete M, Bartels J, Chang S, An G. Translational systems biology of inflammation. PLoS Comput Biol 2008;4(4):e1000014.

[45] An G, Mi Q, Dutta-Moscato J, Solovyev A, Vodovotz Y. Agent-based models in translational systems biology. WIRES 2009;1:159–71.

[46] Clermont G, Bartels J, Kumar R, Constantine G, Vodovotz Y, Chow C. *In silico* design of clinical trials: a method coming of age. Crit Care Med 2004;32(10):2061–70.

[47] Kumar R, Clermont G, Vodovotz Y, Chow CC. The dynamics of acute inflammation. J Theor Biol 2004;230:145–55.

[48] Namas R, Ghuma A, Torres A, Polanco P, Gomez H, Barclay D, et al. An adequately robust early TNF-a response is a hallmark of survival following trauma/hemorrhage. PLoS One 2009;4(12):e8406.

[49] Deitch EA. Gut lymph and lymphatics: a source of factors leading to organ injury and dysfunction. Ann NY Acad Sci 2010;1207(Suppl. 1):E103–11.

[50] Eichacker PQ, Parent C, Kalil A, Esposito C, Cui X, Banks SM, et al. Risk and the efficacy of antiinflammatory agents: retrospective and confirmatory studies of sepsis. Am J Respir Crit Care Med 2002;166(9):1197–205.

[51] Kim M, Christley S, Alverdy JC, Liu D, An G. Immature oxidative stress management as a unifying principle in the pathogenesis of necrotizing enterocolitis: insights from an agent-based model. Surg Infect (Larchmt) 2012;13(1):18–32.

[52] Arciero J, Rubin J, Upperman J, Vodovotz Y, Ermentrout GB. Using a mathematical model to analyze the role of probiotics and inflammation in necrotizing enterocolitis. PLoS One 2010;5:e10066.

[53] Upperman JS, Lugo B, Camerini V, Yotov I, Rubin J, Clermont G, et al. Mathematical modeling in NEC—a new look at an ongoing problem. J Pediatr Surg 2007;42:445–53.

[54] Gopalakrishnan V, Kim M, An G. Using an agent-based model to examine the pathogenesis of surgical site infection. Adv Wound Care 2013;2(9):510–26.

[55] Barie PS. Surgical site infections: epidemiology and prevention. Surg Infect (Larchmt) 2002;3(Suppl. 1):S9–S21.

[56] Cruse P. Wound infection surveillance. Rev Infect Dis 1981;3(4):734–7.

[57] Seal JB, Morowitz M, Zaborina O, An G, Alverdy JC. The molecular Koch's postulates and surgical infection: a view forward. Surgery 2010;147(6):757–65.

[58] Trinchieri G. Cancer and inflammation: an old intuition with rapidly evolving new concepts. Annu Rev Immunol 2012;30:677–706.

[59] Elinav E, Nowarski R, Thaiss CA, Hu B, Jin C, Flavell RA. Inflammation-induced cancer: crosstalk between tumours, immune cells and microorganisms. Nat Rev Cancer 2013;13(11):759–71.

[60] Grivennikov SI, Greten FR, Karin M. Immunity, inflammation, and cancer. Cell 2010;140(6):883–99.

IV. TOOLS AND IMPLEMENTATION OF TRANSLATIONAL SYSTEMS BIOLOGY: THIS IS HOW WE DO IT

[61] Grivennikov SI, Karin M. Inflammation and oncogenesis: a vicious connection. Curr Opin Genet Dev 2010;20(1):65–71.

[62] Hanahan D, Weinberg RA. Hallmarks of cancer: the next generation. Cell 2011;144(5):646–74.

[63] Algra AM, Rothwell PM. Effects of regular aspirin on long-term cancer incidence and metastasis: a systematic comparison of evidence from observational studies versus randomised trials. Lancet Oncol 2012;13(5):518–27.

[64] Rothwell PM, Price JF, Fowkes FG, Zanchetti A, Roncaglioni MC, Tognoni G, et al. Short-term effects of daily aspirin on cancer incidence, mortality, and non-vascular death: analysis of the time course of risks and benefits in 51 randomised controlled trials. Lancet 2012;379(9826):1602–12.

[65] Colotta F, Allavena P, Sica A, Garlanda C, Mantovani A. Cancer-related inflammation, the seventh hallmark of cancer: links to genetic instability. Carcinogenesis 2009;30(7):1073–81.

[66] Zitvogel L, Apetoh L, Ghiringhelli F, Andre F, Tesniere A, Kroemer G. The anticancer immune response: indispensable for therapeutic success? J Clin Invest 2008;118(6):1991–2001.

[67] Nathan C. Points of control in inflammation. Nature 2002;420(6917):846–52.

[68] Nathan C, Ding A. Nonresolving inflammation. Cell 2010;140(6):871–82.

[69] An G, Kulkarni S. An agent-based modeling framework linking inflammation and cancer using evolutionary principles: description of a generative hierarchy for the hallmarks of cancer and developing a bridge between mechanism and epidemiological data. Math Biosci 2014 pii: S0025-5564(14)00142-4. doi: 10.1016/j.mbs.2014.07.009. [Epub ahead of print.]

[70] Hanahan D, Weinberg RA. The hallmarks of cancer. Cell 2000;100(1):57–70.

[71] Chapa J, Bourgo RJ, Greene GL, Kulkarni S, An G. Examining the pathogenesis of breast cancer using a novel agent-based model of mammary ductal epithelium dynamics. PLoS One 2013;8(5):e64091.

[72] Kim SH, Debnath J, Mostov K, Park S, Hunt CA. A computational approach to resolve cell level contributions to early glandular epithelial cancer progression. BMC Syst Biol 2009;3:122.

[73] Engelberg JA, Datta A, Mostov KE, Hunt CA. MDCK cystogenesis driven by cell stabilization within computational analogues. PLoS Comput Biol 2011;7(4):e1002030.

[74] Dutta-Moscato J, Solovyev A, Mi Q, Nishikawa T, Soto-Gutierrez A, Fox IJ, et al. A multiscale agent-based *in silico* model of liver fibrosis progression. Front Bioeng Biotechnol 2014;2:1–10. (Article 18).

[75] Cockrell C, Christley S, An G. Investigation of inflammation and tissue patterning in the gut using a Spatially Explicit General-purpose Model of Enteric Tissue (SEGMEnT). PLoS Comput Biol 2014;10(3):e1003507.

[76] Turner JR. The gastrointestinal tract Kumar V, Abbas AK, Aster JC, Fausto N, editors. Robbins and Cotran: pathological basis of disease (8th ed.). Philadelphia, PA: Saunders-Elsevier; 2004.

[77] Scoville DH, Sato T, He XC, Li L. Current view: intestinal stem cells and signaling. Gastroenterology 2008;134(3):849–64.

[78] Pinto D, Gregorieff A, Begthel H, Clevers H. Canonical Wnt signals are essential for homeostasis of the intestinal epithelium. Genes Dev 2003;17(14):1709–13.

[79] Kuhnert F, Davis CR, Wang HT, Chu P, Lee M, Yuan J, et al. Essential requirement for Wnt signaling in proliferation of adult small intestine and colon revealed by adenoviral expression of Dickkopf-1. Proc Natl Acad Sci USA 2004;101(1):266–71.

[80] Gregorieff A, Clevers H. Wnt signaling in the intestinal epithelium: from endoderm to cancer. Genes Dev 2005;19(8):877–90.

[81] Fevr T, Robine S, Louvard D, Huelsken J. Wnt/beta-catenin is essential for intestinal homeostasis and maintenance of intestinal stem cells. Mol Cell Biol 2007;27(21):7551–9.

[82] Kawano Y, Kypta R. Secreted antagonists of the Wnt signalling pathway. J Cell Sci 2003;116(Pt 13):2627–34.

[83] Miller JA, Baramidze GT, Sheth AP, Fishwick PA, editors. Investigating ontologies for simulation modeling In: Proceedings of the thirty-seventh annual symposium on simulation. Washington, DC: IEEE Computer Society; 2004.

[84] Batts LE, Polk DB, Dubois RN, Kulessa H. Bmp signaling is required for intestinal growth and morphogenesis. Dev Dyn 2006;235(6):1563–70.

[85] Roberts AB, Sporn MB. Transforming growth factor-b Clark RAF, editor. The molecular and cellular biology of wound repair (2nd ed.). New York, NY: Plenum Press; 1996. p. 275–308.

[86] Li NY, Vodovotz Y, Hebda PA, Abbott KV. Biosimulation of inflammation and healing in surgically injured vocal folds. Ann Otol Rhinol Laryngol 2010;119(6):412–23.

[87] Solovyev A, Mi Q, Tzen Y-T, Brienza D, Vodovotz Y. Hybrid equation-/agent-based model of ischemia-induced hyperemia and pressure ulcer formation predicts greater propensity to ulcerate in subjects with spinal cord injury. PLoS Comput Biol 2013;9:e1003070.

[88] Li NY, Verdolini AK, Rosen C, An G, Hebda PA, Vodovotz Y. Translational systems biology and voice pathophysiology. Laryngoscope 2010;120:511–5.

[89] Li NYK, Verdolini K, Clermont G, Mi Q, Hebda PA, Vodovotz Y. A patient-specific *in silico* model of inflammation and healing tested in acute vocal fold injury. PLoS One 2008;3:e2789.

[90] Regan MA, Teasell RW, Wolfe DL, Keast D, Mortenson WB, Aubut JA. A systematic review of therapeutic interventions for pressure ulcers after spinal cord injury. Arch Phys Med Rehabil 2009;90(2):213–31.

[91] Bates-Jensen BM, Guihan M, Garber SL, Chin AS, Burns SP. Characteristics of recurrent pressure ulcers in veterans with spinal cord injury. J Spinal Cord Med 2009;32(1):34.

[92] Bao F, Bailey CS, Gurr KR, Bailey SI, Rosas-Arellano MP, Dekaban GA, et al. Increased oxidative activity in human blood neutrophils and monocytes after spinal cord injury. Exp Neurol 2009;215(2):308–16.

[93] Frost F, Roach MJ, Kushner I, Schreiber P. Inflammatory C-reactive protein and cytokine levels in asymptomatic people with chronic spinal cord injury. Arch Phys Med Rehabil 2005;86(2):312–7.

[94] Segal JL, Gonzales E, Yousefi S, Jamshidipour L, Brunnemann SR. Circulating levels of IL-2 R, ICAM-1, and IL-6 in spinal cord injuries. Arch Phys Med Rehabil 1997;78(1):44–7.

[95] Edlich RF, Winters KL, Woodard CR, Buschbacher RM, Long WB, Gebhart JH, et al. Pressure ulcer prevention. J Long Term Eff Med Implants 2004;14(4):285–304.

[96] Lindblad WJ. Considerations for selecting the correct animal model for dermal wound-healing studies. J Biomater Sci Polym Ed 2008;19(8):1087–96.

[97] Peirce SM, Skalak TC, Rodeheaver GT. Ischemia-reperfusion injury in chronic pressure ulcer formation: a skin model in the rat. Wound Repair Regen 2000;8(1):68–76.

[98] Vodovotz Y. Translational systems biology of inflammation and healing. Wound Repair Regen 2010;18(1):3–7.

[99] Vodovotz Y. At the interface between acute and chronic inflammation: insights from computational modeling Roy S, Sen C, editors. Chronic inflammation: nutritional and therapeutic interventions. Florence, KY: Taylor & Francis; 2013.

[100] Kranke P, Bennett MH, Martyn-St James M, Schnabel A, Debus SE. Hyperbaric oxygen therapy for chronic wounds. Cochrane Database Syst Rev 2012;4:Cd004123.

[101] Tzen YT, Brienza DM, Karg P, Loughlin P. Effects of local cooling on sacral skin perfusion response to pressure: implications for pressure ulcer prevention. J Tissue Viability 2010;19(3):86–97.

[102] Tzen YT, Brienza DM, Karg PE, Loughlin PJ. Effectiveness of local cooling for enhancing tissue ischemia tolerance in people with spinal cord injury. J Spinal Cord Med 2013;36(4):357–64.

[103] An G. Concepts for developing a collaborative *in silico* model of the acute inflammatory response using agent-based modeling. J Crit Care 2006;21(1):105–10. discussion 10–1, PMID: 16616634.

[104] Uschold M, Gruninger M. Ontologies: principles, methods and applications. Knowl Eng Rev 2009;11:93–136.

[105] Consortium TGO Gene ontology: tool for the unification of biology. Nat Genet 2000;25(1):25–9.

[106] Noy NF, Shah NH, Whetzel PL, Dai B, Dorf M, Griffith N, et al. BioPortal: ontologies and integrated data resources at the click of a mouse. Nucleic Acids Res 2009;1(37):170–3.

[107] Rubin DL, Lewis SE, Mungall CJ, Misra S, Westerfield M, Ashburner M, et al. National Center for Biomedical Ontology: advancing biomedicine through structured organization of scientific knowledge. OMICS 2006;10(2):185–98.

[108] An G, Christley S. Agent-based modeling and biomedical ontologies: a roadmap. Wiley Interdiscip Rev Comput Stat 2011;3:343–56. ePublication: April 15, http://dx.doi.org/10.1002/wics.167.

[109] Colasanti R, An G. The Abstracted Biological Computational Unit (ABCU): introduction of a recursive descriptor for multi-scale computational modeling of biologica systems. J Crit Care 2009;24:e35–6.

[110] Solovyev A, Mikheev M, Zhou L, Dutta-Moscato J, Ziraldo C, An G, et al. SPARK: a framework for multi-scale agent-based biomedical modeling. Int J Agent Technol Syst 2011;2(3):18–31.

[111] Railsback SF, Lytinen SL, Jackson SK. Agent-based simulation platforms: review and development recommendations. Simulation 2006;82(9):609–23.

4.5

Getting Science to Scale: Accelerating the Development of Translational Computational Models

Besides black art, there is only automation… **Federico Garcia Lorca**

The most fruitful areas for the growth of the sciences were those which had been neglected as a no-man's land between the various established fields. **Norbert Wiener from "Cybernetics: Of Control and Communication in the Animal and the Machine"**

We have seen in the preceding chapters how computational modeling applied to inflammatory processes can provide useful insights, facilitate the execution of preclinical studies, and used to simulate clinical populations and therapeutic trials. Underlying all these usages is the ability of these methods to accelerate the Scientific Cycle, with the focused goal of enhancing the experimental phase that involves hypothesis testing. The resulting dynamic computational models all represent *in silico* experimental objects that can be manipulated and evaluated in a more rapid time span than either biological proxy models or patient populations, and represent a significant enhancement of the biomedical research community's ability to evaluate, falsify, and refine its mechanistic knowledge. However, as much of an advance in the scientific workflow provided by these *in silico* investigations, the development, construction, and testing of these computational objects remain a laborious, time-consuming, expertise-driven task. This is particularly true since the skill set currently required to generate these types of models is not part of the general biomedical researcher's armamentarium. As a result, a current critical barrier to the general adoption of dynamic computational modeling by the biomedical research community is the actual process of constructing a model.

What do we mean by this statement? There are two components to our answer: the first involves a cognitive and philosophical barrier, while the second represents a logistical one. The cognitive and philosophical barrier has to do with the intellectual task of creating a model, i.e., creating a useful abstraction. We have shown in our review of the history of biological research and its current manifestation in the research community that, given a legacy that equates knowledge with detail, being able to represent their knowledge in the form of useful abstractions is not something they are trained to do. This is why simply providing bioscientists better tools to do modeling will not address this issue: this barrier *cannot* be overcome by simply providing biomedical researchers an improved user interface, a software toolkit, or a new modeling language. A "model" in bioscience, as we have seen, is generally thought of as a highly engineered biological proxy, with an operational emphasis on getting those models to perform to predetermined specifications, i.e., be "validated." The knowledge extracted from those models is represented in schematics and diagrams (i.e., the proverbial "Slide #3" in a bioscience talk, following the Title Slide #1 and the Hypothesis Slide #2 with a slide showing a diagram of the pathway/process of interest). Despite their reliance on these diagrams, bioscientists do not generally recognize the potential power associated with these visualized hypotheses; rather most times they are presented apologetically, with numerous statements about how of course the real system is much more complex. This attitude incorporates a false promise: that if the diagram was just more comprehensively filled out, then "the answer" would be achieved. However, as we have seen, this is the path to infinite regress, and misses the opportunity provided by having at least a cartoon of the their knowledge. This is because, to a great degree, that cartoon represents what that researcher considers to be the "essentials" of their system of study (of course, or else they wouldn't be studying it!). As such, you would think it would behoove that researcher to identify just how powerful their conceptual model might be: how much biology can this "model" actually explain? What is the range of its impact? Let us pursue this cartoon analogy just a little bit more: a cartoon is often also an animation, which implies motion and behavior. As we have seen in our chapter on Dynamic Knowledge Representation, the essence of biology is that it involves dynamics; what we need to do is make biological researchers the benefits of being able to make their hypotheses "go." Once bioscientists have these dynamic abstractions, then they try and use them to explain the different types of observations they collect. To a great degree, this flips the board in terms of the view of experimental variability. Where once (i.e., now) the goal is to limit experimental variability in order to get "clean" data that can pass statistical tests, now (i.e., the future) goal would be to see what range of biological heterogeneity could conceivably be explained using a dynamic instantiation of a conceptual model. This process is the pathway toward theory, and a deeper understanding of the biology that binds things together as opposed to make them distinct.

But, regardless of the intellectual appeal of this scenario, given the status quo in biological research biologists will need convincing in order to shift their worldview. The natural way to accomplish this is through demonstrations and examples of how this process would work. Unfortunately, this requires that such examples be generated, not only in a general sense (see all the examples provided in Chapters 4.2–4.4), but within their personal or near-personal experience. It is at this point that we run into the second, now logistical, barrier to the greater use of computational modeling: getting models made. Currently, computational bioresearch involves a biologist, who knows the particulars of the biology being studied and the type of research questions that need to be asked, interacting with a computational modeler, who is often an expert in a particular type or set of modeling methods and can provide information about the capabilities of those methods. There is often little pre-existing intersection in mutual areas of expertise; those that have familiarity in both areas are an even more rare resource. A conversation between these experts is necessary to create a mapping between the biological knowledge and the possible computational methods to identify how components, aspects, or properties of the biological system can be suitably represented with what set of mathematical formalisms and for a given desired purpose. As a result, a bioscientist embarking on this path very often requires finding and establishing a collaboration with experts in applied mathematics, computer science, or engineering, a step that we have already seen runs up against the compartmentalized, siloed nature of the scientific community. The fact that computational bioresearch requires multidisciplinary and multidomain expertise, and that the initiation of such a collaborative relationship is a labor intensive and time consuming collaborative endeavors, presents significant practical logistical barriers that limit the permeation of computational approaches into the wider research community. This situation leads us to recognize that we need to be able to scale our strategies for navigating the Scientific Cycle in order to meet the demands of a data-rich, high-throughput world. To find our answer we turn, naturally, to technology.

The past few decades has seen increasing research in computation-aided discovery of scientific knowledge [1]. While originally envisioned as the development of intelligent computer systems that could produce new scientific discoveries on par with a human scientist [2], the field has grown to encompass creativity processes and computational augmentation of the Scientific Method, such as generation and evaluation of models, design of experiments, management of scientific knowledge, and construction of scientific workflows. Most computational and automated discovery research has concentrated on the manipulation of well-defined formal structures such as mathematical statements, mathematical equations, and logic programs. Research in this process of automated inference has resulted in the advancement of artificial intelligence (AI) algorithms to perform deductive, inductive, and abductive reasoning.

All these developments are fine, but we believe that there is still an aspect of human intuition and genius that cannot yet be reproduced with an algorithm (but we do think we humans might get a little help!) Therefore, rather than trying to completely automate the discovery process, what we would like to do is automate around the human capacity to exercise intuition and expertise, augmenting those steps that can be treated algorithmically while still continuing to take advantage of currently irreplaceable human insight and keep the human researcher in the discovery loop. Doing so should be aimed at enhancing the ability of researchers to accelerate the Scientific Cycle, in which observations are converted to hypotheses, hypotheses are tested through experiment, and the results of experiments are evaluated to refine our knowledge. Since humans store and communicate knowledge through natural language, we suggest that effective strategies need to be able to process natural language content. This capability will leverage both the repository of knowledge found in the literature as well as facilitate the ability of researchers to express their own hypotheses. Since there are also many methods of generating putative hypothesis from large data sets, any possible solution should also incorporate the ability to accept, as input, such types of knowledge structures. Also, since very often the type of computational model is constrained by the type of data available to construct it, any potential solution will also interface with existing knowledge repositories and databases.

Below, we describe one example of this approach. This is not intended to be a comprehensive or unique solution to the Translational Dilemma, but it does represent an initial attempt to engineer a robust and scalable "cyberinfrastructural" solution that addresses the issues that we have identified. This computational investigatory framework is centered around an AI software agent, the Computational Modeling Assistant (CMA), that can fill the critical link between biological knowledge and computational models, enhance the ability of basic researchers to represent their own hypotheses, and to explore the putative dynamic behavior of those hypotheses in an *in silico* environment [3]. The CMA is designed to augment the ability of bioscientists to represent their knowledge in computational form by utilizing algorithms to address repetitive portions of the model construction workflow, thereby allowing them to more readily perform *in silico* experiments that will in turn guide future real-world experiments. The structure of a CMA-augmented discovery workflow is seen in Figure 4.5.1. The CMA leverages the increased capabilities for gathering the data needed to create a model (ontologies, text analysis, and automated inference) with advances in the development of model markup languages, toolkits, and modeling standards. The CMA accelerates the modeling process by treating the mapping between biology and computation as an AI planning task to identify suitable modeling methods and create the selected models. The CMA leverages the work done in the systems biology community related to the representation and sharing of computational biology models in Systems Biology Markup Language (SBML) [4]/Cell Markup Language (CellML) [5], but extends modeling capability to address many other questions, such as those concerned with spatial effects, structural properties, biomechanics, population behavior,

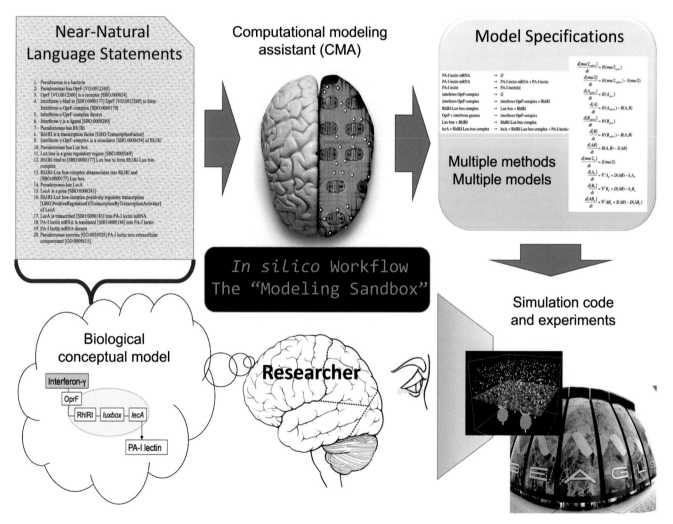

FIGURE 4.5.1 **Schematic role of the CMA.** The CMA serves to augment the Scientific Cycle by facilitating the translation of biological knowledge/hypotheses into dynamic computational models, which can be executed to identify which model structures exhibit plausible behavior. The greatly increased ability to evaluate hypotheses *in silico* will help remove the current bottleneck in the existing data-rich environment (see Figure 2.3.1).

and macro-scale output (such as tissue and organ physiology), that may require mathematical modeling methods currently not expressible in those formats. The CMA provides these capabilities to the bioresearcher without the requirement of mathematical expertise, by allowing the researcher to express their hypotheses using natural language. The potential benefits of this framework cross-multiple disciplines: tools such as the CMA would facilitate the entry of basic researchers into computational modeling and allow them to evaluate their hypotheses through the design of new experiments, and the CMA will benefit computational researchers by improving scalability with respect to the scope of model sharing and interoperability projects. Finally, the CMA would facilitate collaborations between biologists and computational scientists by quickly bridging the cognitive gap between biological knowledge and mathematical concepts, and rapidly producing prototype computational models that focus on the significant causal mechanisms underlying the biological phenomena.

As we have noted, the goal of attempting to use computational methods to integrate and advance biomedical knowledge has received a great deal of attention and effort. However, as we have also seen repeatedly, Bacon's Idol of the Marketplace and the ambiguity of language returns again, since the terminology used to describe the various approaches biomedical knowledge integration and facilitated model construction can be quite ambiguous. Therefore, in order to help clarify what the CMA actually is, we start by describing what CMA is not:

1. The CMA is not a *database*. Databases are collections of information and knowledge; CMA executes a process aimed at converting a specific hypothesis (which may be informed by databases) into a specific set of computational modeling (which may also be informed by databases). The key here is that the CMA is an active computational agent that executes a process, not just a mere repository. However, the CMA does contain a KnowledgeBase that the AI agent uses as an interface for various data and knowledge sources.
2. The CMA is not a *modeling language*. The CMA uses natural language as a means of representing biological knowledge, and does not require the biologist to learn any new words or format with which to express their knowledge. However, the artificially intelligent software agent within the CMA does use an internal logical language for reasoning about biological and computational concepts.
3. The CMA is not a *toolkit*. Modeling toolkits are generally intended to facilitate a specific type of mapping between specific types of biological knowledge (such as molecular pathways) and specific types of modeling methods (such as ordinary differential equations). If there are established markup languages, these might be an input format to a modeling toolkit. A modeling toolkit has intrinsic limitations in its expressiveness, because of its focus on specific types of biological knowledge and modeling methods, thus restricting the fidelity of the mapping between biology and computational model as a biologist seeks to recast their knowledge into the toolkit's format. However, the CMA can utilize modeling toolkits for executing simulations if the computational model aligns with the capabilities of the toolkit.
4. The CMA is not a *markup language*. The CMA is not a means of expressing knowledge, it is a process engine that converts knowledge presented in one way (natural language for biology) into another (formalisms for mathematical/computational modeling methods). The CMA can convert biological statements of an appropriate structure into existing markup language that are in use, such as SBML and CellML, but also has the expressiveness to represent biological concepts not currently supported with those markup languages, such as spatial configurations, structural properties, biomechanical interactions, population effects and higher order, macrolevel behaviors (such as tissue, organ, and multiorgan physiology). However, the CMA does employ markup languages for structuring the knowledge and data that it uses and generates as part of the mapping and modeling process.
5. The CMA is not a *model repository*. The CMA does not store all models for any specific biological phenomena, rather it is a process engine to generate the models that fill a repository. However, the CMA does contain a model management system that provides provenance for computational models and *in silico* experiments.
6. The CMA is not a specific *project* on model sharing or interoperability. The CMA is indirectly related to work on model interoperability, not by trying to integrate existing models; rather, the CMA could suggest different modeling approaches applied to different components of a particular biological system (i.e., hybrid models). The CMA also provides alternative modeling methods to represent the same biological hypothesis.

Tools such as the CMA would aid in the democratization of science by increasing the scope and scale of hypothesis generation, communication, evaluation, and refinement for the community at large. These types of cyberinfrastructures are predicated upon developing a wide range of interactions for: natural language processing, hypothesis generation from data, specific modeling methods, knowledge bases, and a test-bed wet lab investigatory workflow. Development of these types of systems requires incorporating specific examples of these various interfaces, but with the recognition that each of these areas is a topic for rich ongoing development and the goal of this proposal is to be able to leverage developments in these different areas. The goal of this type of computational framework is meta-scientific and not focused on targeting a particular question in a particular aspect of science: it is about addressing the *scientific process* itself.

THE STRUCTURE OF THE CMA

Let us first examine the components needed for an automated tool such as the CMA to carry out its intended function (Figure 4.5.2). The CMA utilizes and leverages developments in formal knowledge representation in biology and modeling and simulation with an intelligent agent approach of automated reasoning to aid in the construction of a computational model specification and simulation code. The biological content of the KnowledgeBase available to the CMA includes current knowledge extracted from the literature, computational predictions from bioinformatics analysis, and investigator-derived hypotheses. This KnowledgeBase would have its set of biological entities, processes, and functions annotated to corresponding biomedical ontological terms, thus providing a controlled vocabulary and structure to the knowledge. The CMA utilizes a set of rules that encapsulates expert knowledge of modelers concerning the process of computational model construction. These rules represent specific linkages between the biological system being modeled and the modeling methods available to represent those types of systems. To utilize a tool such as the CMA, a researcher hypothesizes a conceptual biological model by selecting a subset of biological entities, functions, and relationships from the KnowledgeBase this conceptual biological model is passed to the CMA, which then treats model construction as a planning task where the goal is translating the researcher's conceptual biological model into simulation code. The plan is divided into two parts: constructing the model specification and producing the simulation code.

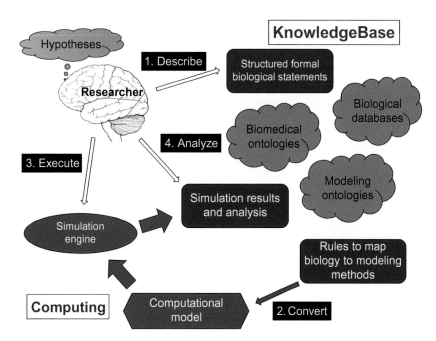

FIGURE 4.5.2 **Various components and workflow of the CMA.** A researcher has knowledge that is constructed into a hypothesis. The researcher then expresses that knowledge in a series of near-natural language statements/simple sentences (Step 1: *Describe*). The CMA then processes those biological statements using its KnowledgeBase, which includes structured biological knowledge and terminology from biomedical ontologies, data present both in local and generally available databases, and known structures/specifications of different types of modeling methods (modeling ontologies). The KnowledgeBase also includes known rules for mapping biological processes into different types of modeling methods (the step that incorporates the cataloged model-making knowledge of expert computational modelers), and these are used to generate a set of computational models (Step 2: *Convert*). The resulting computational models are then executed using a simulation engine to generate simulation data, i.e., the performance of simulation experiments (Step 3: *Execute*). The resulting simulation data is then re-introduced into the CMA's KnowledgeBase for analysis, including comparison against existing observations and data present (Step 4: *Analyze*). Note that the Researcher takes an active role at various steps in the workflow, keeping the human, with their expertise and intuition, in the investigatory loop.

The CMA could translate a set of formal biological statements stored in the KnowledgeBase into one or more computational model specifications that can be simulated either by using existing simulation toolkits or directly generating executable simulation code. The CMA as instantiated currently has six primary components:

1. The KnowledgeBase of biological concepts, computational modeling concepts, and numerical methods
2. Rules for mapping biological concepts into modeling methods
3. Rules for mapping modeling methods into computer code
4. A logic inference engine
5. An intelligent planning agent
6. A model management system.

The CMA will have a modular design with well-defined interfaces for each component thus allowing the components to evolve separately in their internal design and implementation. This is a key requirement for effective scalability and broader applicability because the CMA is intended to be agnostic to the specific biological phenomena being studied and to any specific modeling method.

The conversion of a biological conceptual model into a computational model specification is treated as a planning task. The intelligent agent (i) defines an initial state from a given set of formal biological statements (subset of literature-extracted information, bioinformatics analysis, researcher-provided hypotheses in the KnowledgeBase or from external sources described below), (ii) defines one or more goal states for desired model specifications, and (iii) utilizes the logic inference engine to apply mapping rules for transition from one state to the next. The sequence of rules from initial state to goal state is the resultant plan for modeling the biology, and an actual model specification is generated in the process, which can then be executed using existing or custom simulator environments. The CMA intelligent software agent is implemented in the Maude logical framework [6] that constructs an ODE/PDE, Petri net or agent-based model specification for a given biological conceptual model.

This last point represents the novel impact of a tool such as the CMA: it is not tied to any specific modeling method, and multiple goal states correspond to multiple model specifications, which are presented as set of possible alternative modeling methods. Rather, tools such as the CMA would provide guidance for the best modeling choice that addresses the user's primary interest and can be validated with the type of experimental data available. More than one model specification can be chosen, thus allowing comparison of the different models which can help point out whether a simpler model is sufficient or whether a hybrid model combining multiple methods is required to properly model the biological system. The importance of this novel capability of CMA is reinforced by the growing recognition that different components or modules of a biological model may be best represented with different simulation methods. For example, one specification might be for a continuous deterministic model, while another may be a discrete stochastic model. The researcher reviews the presented model specifications and chooses one or more for CMA to produce simulation code. This decision is based upon various factors specific to the conditions in which the model construction task is being performed: availability of certain types of biological data, prior knowledge about the behavior of the biological system, and level of modeling detail. Only the researcher has the awareness and intuition to assess the experimental context of the model construction task and possess the ability to adjudicate between the usefulness of one model abstraction versus another based on the biological question at hand. Furthermore, since the plans developed by CMA are transparent, an explanatory description of how the biology is modeled is provided. Even if a complete model specification cannot be obtained, the intelligent agent can indicate which biology statements cannot be transformed. These capabilities allow the CMA to be viewed as a tool that will facilitate interdisciplinary communications and enhance the process of forming collaborations.

KNOWLEDGE BASES IN THE CMA

An automated tool should facilitate interaction with the published literature. The CMA's KnowledgeBase includes interfaces with biomedical ontologies currently found in an online repository, BioPortal (http://www.bioportal.bioontology.org), which is managed by the National Center for Biomedical Ontologies (NCBO) (http://www.bioontology.org). The NCBO has ongoing development of computational and bioinformatics tools to increase the utility of bio-ontologies for the biomedical community at large, and the CMA utilizes these tools (BioPoral REST services) when possible to leverage their extensive capabilities in terms of searching across different ontologies, identifying relationships between terms, and utilizing their structured vocabulary. Additionally, the OBO Foundry (http://www.obofoundry.org) is a collaborative project that is focused on establishing principles for biomedical ontology development with the goal of growing a group of interoperable ontologies to be used in the biomedical arena; a substantial number of the ontologies found in BioPortal originate from the OBO Foundry. The CMA conforms to the OBO Foundry design principles with respect to the development and representation of its KnowledgeBase of biological concepts. On the modeling method side, ontologies for computational modeling concepts are not as developed as their biomedical counterparts. Markup languages like OpenMath and MathML provide for mathematical expression, while traditional ontologies like DeMO [7–9], Ontology for Physics in Biology [10], and KiSAO [11] are also utilized.

The CMA is also able to produce simulation code that can be executed from the model specification. In some cases, the model specification aligns with an existing modeling markup language like SBML [4] or CellML [12] thus allowing existing simulation toolkits such as COPASI [13] or CellDesigner [14] to be utilized to execute simulations. Even model specifications that go beyond molecular pathways and include spatial organization and cellular behavior may be simulated with toolkits providing some of that capability such as CompuCell [15]. However, currently there is no simulation toolkit that is general enough to simulate every possible computational model, and therefore the CMA can directly produce the simulation code to be executed. The conversion of a model specification into simulation code is similar to the generation of the model specification in that it is treated as a planning task. The initial state is the computational model specification, and the goal state is a set of programming language code segments that implements the model. The sequence of rules from initial state to goal state is the resultant plan for implementing the model specification, and actual computer source code is generated in the process, which might be generated for a particular programming language or software toolkit. Here, CMA utilizes the ontology of numerical methods along with a set of mapping rules to transform the model specification into simulation code using specific numerical methods.

Just as with construction of the model specification, multiple numerical methods can be applied for the simulation of the same model. These alternatives are typically concerned with computational performance and stability issues. They play less of a role regarding interpretation and analysis of the biological model, but they are important for insuring the simulation produces valid results. For example, a Runge–Kutta fourth-order method for solving ordinary differential equations (ODEs) could provide greater stability and a larger time step over the basic Euler method for solving ODEs, though it may have a greater computational cost. The CMA might be able to analyze the parameter values for the model, looking for order of magnitude scale differences that indicate a stiff system, thus requiring more stable methods to be used. Furthermore, the CMA can take advantage of information about the hardware, thus producing simulation code that is geared toward a specific computing environment whether it be a single computer, multicore processor, parallel Message Passing Interface cluster (the dominant structure of today's massively parallel supercomputers), or a Graphics Processing Unit (GPU), which are increasingly used for high-performance computing. While this scalability across hardware platforms is not gained for free, as the knowledge base of numerical and simulation methods increases over time more of this capability can be provided in an automated fashion.

Besides the user's data, a tool such as the CMA would have the capability to query external data sources. These sources might provide data such as kinetic rates for parameter values in the model or additional observations for calibrating and validating the model. More importantly, data-driven discovery techniques (see Chapter 4.2) may be employed to generate numerous putative hypotheses, which can be used as biological knowledge inputs for the CMA. In keeping with our modular design, our goal is not to perform data integration of heterogeneous sources, nor to write specialized interfaces for each specific source. Instead, a tool such as the CMA would define a query interface for generic statements posed by the AI, for example, "what is binding rate for a protein to a DNA site," "what transcription factors regulate a gene," or "what gene expression data is available for a gene." Bioinformatics analysis of new experimental data will yield new hypotheses that will either require further experimental data or model development to determine the appropriate testing forum.

Rules for Mapping Biological Concepts into Modeling Methods

The CMA's mapping rules encapsulate the expert knowledge of modelers and the methods for representing biological concepts with computational models. These rules allow the CMA to perform the initial steps taken by a trained computational modeler. The rules are specified as equation-based rules in conditional rewriting logic [16] stored in the KnowledgeBase in a general format. Our existing rule set includes biological entities such as genes, mRNA, proteins, protein complexes, cells, bacteria, ligands, receptors, and promoter regions; and includes biological processes such as transcription, translation, degradation, binding, disassociation, enzymatic activity, regulation, inhibition, activation, secretion, and diffusion. These rules currently support mapping those biological concepts into a number of modeling methods including ODEs, PDEs, Petri nets, and agent-based models.

One of the challenges we identified is the construction of hybrid models whereby disjoint portions of the biology are represented with different modeling methods, for example, discrete agent-based cells with PDE of spatially diffusing molecules or an intracellular stochastic gene network coupled with agent-based rules for cell movement and division. There are two aspects to be considered. One is the definition of sets of biological concepts that would typically be translated together into the same modeling method, and the other is the specific coupling technique to interface the different methods. Without guidance for the first aspect, mapping rules can arbitrarily translate individual biological concepts into different methods leading to a combinatorial explosion of "hybrid" models. A tool such as the CMA addresses this by defining generic rules for typical groups of biological concepts (e.g., an intracellular gene network) that would eliminate the combinatorial explosion of potential models by "pruning" paths in the planning task. Likewise, the CMA provides the capability for the user to define groupings of modeling methods that categorize and define the coupling techniques for the modeling methods supported by CMA. These coupling techniques are themselves modeling methods with annotations for the matching of variables and functions, physical units, and time scales. By treating these mapping rules as knowledge structures in themselves that can be independently developed, this provides a broader impact whereby future research can be targeted at advancing the CMA's capabilities and expressiveness while maintaining interoperability with established but ongoing development in the areas of formal semantics/knowledge representation and modeling methods.

The Logic Inference Engine for the CMA

The CMA uses Maude [6] as its logic inference engine because its rewriting logic is simple yet expressive [16]. Maude provides capabilities such as reflection (high-order logic), model checking, and inference strategies beyond traditional logical frameworks such as CLIPS or Prolog. These capabilities effectively allow the CMA scale up and handle large biological conceptual models, a greater array of computational modeling methods, and the construction of hybrid multiscale computational models. The CMA performs reasoning on the constructed model specifications to extract attributes such as parameters and input variables, and also to discover dynamic behavior and properties without the need to run simulations. For example, the CMA can inform the user if the model is incomplete because some variable will grow without bound (such as when production terms are not matched with decay terms) and suggest inclusion of additional statements from the KnowledgeBase. This desirable capability has a broader scientific impact as it enables researchers to extract higher level insights about computational models themselves as the focus of investigation, beyond just analysis of simulation results.

Model Management System

The workflow through a tool such as the CMA involves multiple steps in which the researcher interfaces with the system and makes selections among presented options. Given the nature of scientific investigation, it is expected that the researcher would need to backtrack along that process and pursue alternative pathways. As such, the CMA has a model management system that will maintain provenance data about each step in the knowledge integration task, from the initial literature used to enrich the researcher's hypothesis, through the various formal representations of knowledge and model specifications, to the simulation experiments and the data used to execute them. This allows tracking of the directions of a particular investigatory path, leading to a comprehensive description of the exploration space related to a particular project.

FULFILLING THE GOALS OF TRANSLATIONAL SYSTEMS BIOLOGY AND THE DEMOCRATIZATION OF BIOMEDICAL SCIENCE

The purpose of the CMA is to enhance the ability of research scientists to do *in silico* science and facilitate the goals of Translational Systems Biology. Translational Systems Biology is predicated on enhancing the ability to instantiate, execute, evaluate, and manipulate hypotheses *in silico*, the critical step in addressing the throughput issues associated with the need to shift through a very large set of putative hypotheses before performing real-world experiments. We have already noted the fact that computational modeling expertise is currently, and for the foreseeable future, a scare resource when compared to the number of potential applications of that expertise. We also recognize that the willingness of biologists to adopt computational methods on their own nearly always falls short of desired expectations. The rationale for the development of cyberstructures such as the CMA directly addresses both of these issues, but there is also an additional benefit from our research strategy of focusing on the mapping/modeling process in terms of training the next generation of researchers. We note that the mapping rules incorporated into the CMA, as well as the process for how the KnowledgeBase is used to inform modeling decisions, are basic modeling principles taught to prospective computational modelers. Examination and familiarity with the inner workings of the CMA represent a practical course in teaching computational modeling, and its repetitive use serves a pedagogical role for biologists, both currently active and especially those in training. Furthermore, the mapping accomplished by the CMA need not be unidirectional; by seeing how biology-to-computation mappings occur can provide computational modelers insight into the types of biology and biological questions most suited to their methods. CMA mappings may also provide insight into novel pathways for computational model transformations that use biological knowledge structures links between previously unconnected methods. We believe that by formally targeting the mapping and model-making process CMA offers a paradigm shifting, transformative opportunity to introduce an entirely new area of scientific investigation that will help to break down the various barriers between disciplines and foster the sharing of knowledge and skills within a collaborative cyberenvironment.

We believe that it is the development of this new generation of multidiscipline capable researchers that will eventually lead to dealing with the Translational Dilemma, and we also believe that they will do so within cyberenvironments like that prototyped with the CMA. Our running theme throughout the book has been to make a foundational analysis of the scientific process questions underlying the Translational Dilemma: Translational Systems Biology is the outcome of that investigation. Intrinsic to this field is the understanding that we also need to implement the tenets of our putative solution, and therefore involves performing process engineering in order to be able to have sustainable, robust, and scalable implementation with which to deal with the complexity of disease. The specifications underlying the CMA represent the result of that exercise. We are hopeful that integrating technologies like the CMA will form the basis of a more open, less rigid scientific landscape, where merit in terms of innovation and actionable success will overcome some of the entrenched shibboleths in the biomedical research community. We will see in our final section how the multiple components of Translational Systems Biology can be brought together and provide a true path forward for our quest to improve human health.

References

[1] Džeroski S, Langley P, Todorovski L. Computational discovery of scientific knowledge Lecture notes in computer science, vol. 4660. Berlin/Heidelberg: Springer; 2007. p. 1–14.

[2] Langley P. The computational support of scientific discovery. Int J Hum Comput Stud 2000;53(3):393–410.

[3] Christley S, An G. A proposal for augmenting biological model construction with a semi-intelligent computational modeling assistant. Comput Math Organ Theory 2012;18(4):380–403.

[4] Hucka M, Finney A, Sauro HM, Bolouri H, Doyle JC, Kitano H, et al. The systems biology markup language (SBML): a medium for representation and exchange of biochemical network models. Bioinformatics 2003;19(4):524–31.

[5] Schilstra MJ, Li L, Matthews J, Finney A, Hucka M, Le Novere N. CellML2SBML: conversion of CellML into SBML. Bioinformatics 2006;22(8):1018–20.

[6] Clavel M, Duran F, Eker S, Lincoln P, Marti-Oliet N, Meseguer J, et al. All about Maude—a high-performance logical framework. Berlin/Heidelberg: Springer-Verlag; 2007.

[7] Fishwick PA, Miller JA. Ontologies for modeling and simulation: issues and approaches. Simulation conference, 2004 proceedings of the 2004 Winter 2004;1: 264. doi:101109/WSC20041371324.

[8] Investigating ontologies for simulation modelingMiller JA, Baramidze GT, Sheth AP, Fishwick PA, editors. Proceedings of the thirty-seventh annual symposium on simulation. Washington, DC: IEEE Computer Society; 2004.

[9] Silver GA, Miller JA, Hybinette M, Baramidze G, York WS. DeMO: an ontology for discrete-event modeling and simulation. Simulation.

[10] Cook DL, Mejino JLV, Neal ML, Gennari JH. Bridging biological ontologies and biosimulation: the ontology of physics for biology. AMIA Annu Symp Proc 2008:136–40.

[11] Courtot M, Juty N, Knüpfer C, Waltemath D, Zhukova A, Dräger A, et al. Controlled vocabularies and semantics in systems biology. Mol Syst Biol 2011;7:543.

[12] Cuellar AA, Lloyd CM, Nielsen PF, Bullivant DP, Nickerson DP, Hunter PJ. An overview of CellML 1.1, a biological model description language. Simulation 2003;79(12):740–7.

[13] Hoops S, Sahle S, Gauges R, Lee C, Pahle J, Simus N, et al. COPASI—a COmplex PAthway SImulator. Bioinformatics 2006;22(24):3067–74.

[14] Matsuoka Y, Funahashi A, Ghosh S, Kitano H. Modeling and simulation using CellDesigner. Methods Mol Biol 2014;1164:121–45.

[15] Izaguirre JA, Chaturvedi R, Huang C, Cickovski T, Coffland J, Thomas G, et al. CompuCell, a multi-model framework for simulation of morphogenesis. Bioinformatics 2004;20(7):1129–37.

[16] Meseguer J. Conditional rewriting logic as a unified model of concurrency. Theor Comput Sci 1992;96:73–155.

SECTION V

A NEW SCIENTIFIC CYCLE FOR TRANSLATIONAL RESEARCH AND HEALTH-CARE DELIVERY

5.1

What Is Old Is New Again: The Scientific Cycle in the Twenty-First Century and Beyond

What's past is prologue. **William Shakespeare, The Tempest**

...Now this is not the end. It is not even the beginning of the end. But it is, perhaps, the end of the beginning. **Winston Churchill**

In the midst of chaos, there is also opportunity. **Sun Tzu**

EVERYTHING OLD IS NEW AGAIN

We hope that by this stage we have made the case for why the Translational Dilemma, the inability to translate basic science knowledge effectively and efficiently into clinical therapeutics, is the greatest challenge facing the biomedical community today. We have examined the historical, operational, and social factors that underlie the Translational Dilemma, and have come to the conclusion that a current imbalance and fragmentation of the Scientific Cycle and its implementation, which paradoxically was driven by initial successes using a reductionist, compartmentalized and specialist-oriented paradigm, is the root cause of the translational challenge. As a result of our analysis, we have proposed, developed, and outlined how Translational Systems Biology holds promise with regard to removing key barriers in this fragmented continuum.

We appreciate the fact that this is a bold statement. As we conclude this book, we would like to make an even bolder statement: rather than proposing a "new kind of science" we present a call for the return to, and a reinvigoration and modernization of, the Scientific Method via the adoption of Translational Systems Biology. By this we mean the incorporation of computational modeling at all phases of the Scientific Cycle in biomedical research, with the tenets of Translational Systems Biology as guiding principles. We make the ambitious assertion that in order to move beyond the current bottleneck in the therapy development pipeline, the adoption of a strategy that mirrors Translational Systems Biology is not only a possible solution, but a *necessary* one. Unfortunately, as we have established in this book, there are considerable barriers to such a shift, heavily reinforced by habit, psychology, cognitive prejudices, and a socio-professional-economic incentive structure that has fragmented the metrics of "success" away from the ostensible goal of biomedical research: the quest to enhance human health. To a great degree, the challenges of the Translational Dilemma do not stop with the "under the skin biology" of determining whether a particular drug will work. In order to truly propose a strategy to overcome the Translational Dilemma, we must also examine how such a strategy can be implemented within the socio-professional-economic community context that established the conditions that produced and currently sustain the Translational Dilemma. As with the "systems diseases" that prove to be so difficult to modulate, so, too, is the biomedical research "system" a multiscale web of mechanisms, causal factors, and feedback interactions that act to sustain the system, even in a disordered and dysfunctional state. Therefore, as with any attempt to modulate a multiscale, complex system, our strategy of Translational Systems Biology must target multiple different control points at multiple different levels, ranging from attempting to overcome individual cognitive prejudices by (re)educating researchers as to the rationale behind their current practices, to providing operational tools to aid in their current investigations, to developing unifying technologies that can ease the reintegration of the fragmented research community, to proposing potential social and economic disruptions that are often needed to transform embedded and tradition-bound socio-professional-economic structures. In this final chapter, we will summarize the different levels targeted by Translational Systems Biology and present some scenarios by which this transformation might eventually play out.

"DATA" IS NOT THE ANSWER; KNOWLEDGE IS

From an individual cognitive awareness level, in order to solve a problem one must first understand that the problem exists, and from whence it arises. It should be clear by now that biomedical research today faces a critical dilemma. As greater detailed information becomes available for increasingly finer levels of biological mechanism, it has become more and more difficult to translate basic mechanistic knowledge into clinically effective diagnostics and therapeutics. We have argued that this Translational Dilemma is most evident in attempts to understand and modulate "systems" processes/disorders, such as sepsis, cancer, autoimmune disease, obesity, diabetes, and wound healing. The complexity of these diseases and their underlying biology has led to a realization that traditional reductionist methods, i.e., the utilization of sequential, single-variable "good" experiments, is insufficient for their effective characterization. But does that mean that the iterative cycle of the Scientific Method—observation, hypothesis generation, hypothesis evaluation via experiment and subsequent hypothesis refinement—is no longer valid? Certain trends in biomedical research would appear to promulgate this view. "Data"—both the term itself as well as actual data—is now used with reverence approaching religious fervor, and "mining" those data is viewed as the key to revelation. Holders of this dogma might say: "We hope that if only enough data could be collected about a system, then a pattern recognition algorithm would somehow produce *the* answer." Unfortunately, this sounds suspiciously like magical thinking. Therefore, the question must be raised as to whether people actually believe that with sufficient data, correlation can be conflated with causality, and that therefore high volume data analysis represents the future of science; if so, then we are concerned that this is not actually the future, but rather harkens to the pre-scientific past. If Big Data is not the answer, or at least not the whole answer, then what is?

In this book, we have tried to outline the development, success, and subsequent derailment of the Scientific Method. We have put forth the proposition that the currently perceived condition is due, in great degree, to imbalances in the iterative loop of the Scientific Cycle. We have tried to outline the technological, political, and economic factors that have helped drive the differential advances between some aspects of the biomedical research process as opposed to others. The initial portion of this book is, in essence, a "root cause analysis" of the current translational dilemma, consisting of a series of assertions regarding biomedical science, and then leading to a proposed approach for addressing the translational challenge within the context of the Scientific Method. We hope that we have convinced you, the reader, that:

- *Assertion 1*: Biological knowledge will always be incomplete
- *Assertion 2*: The goal of translational biomedical research is to be able to intervene in a pathophysiological process, and that goal requires an inference of mechanistic causality in the process being targeted
- *Assertion 3*: The complexity of pathophysiological processes being studied today and for the foreseeable future precludes the use of purely experimental methods for the evaluation of multiscale causality, especially if there is any expectation of translating to the clinic insights derived from basic science studies
- *Assertion 4*: Data-driven computational analysis alone cannot be used to evaluate causality, though this type of analysis can be the initial step on the road to creating computational mechanistic models; because
- *Assertion 5:* Dynamic mechanistic models, which are computational instantiations of mechanistic hypotheses, are the only rational, cost-effective solution for high-throughput multiscale hypothesis evaluation.

We have made the point that the goal of biomedical research should not be that of completeness of description but rather that of sufficiency for practical utility. We have attempted to show multiple examples from our work in which computational models, built on much less than complete data or mechanisms, were capable of yielding novel, potentially clinically useful insights in the context of acute inflammation and critical illness. We believe that biomedical researchers (and the funding agencies that fund them and the policymakers that control their budgets) should not conflate the current lack of effective therapeutics or accurate diagnostics arising directly from our work with the fact biomedical research community has not yet found the "magic bullet" or the next mediator or pathway. Note that this emphasis on clinical translation does not preclude "discovery for discovery's sake;" even this goal requires the researcher to make a decision about sufficiency, if only to answer the question "How much do I need to know to be able to make my next experiment?" From a translational standpoint, this utility can be determined by asking the question: "How do I know enough to be able to effectively modulate this disease process?"

If we depart from this starting point, then we must acknowledge that biological systems operate on mechanisms representing cause and effect, and that these mechanisms can be modulated in a rational fashion. However, since Assertion 1 states that any presumed causality cannot be completely known, we must have some way of establishing "trust" in a particular mechanism. Accomplishing trust in causality is the goal of the Scientific Method.

We have tried to make the case that the biomedical research community, from graduate students working on basic biology problems to scientists in industry, has focused its efforts on ever more detailed, finer grained, and highly manipulated experimental models in the search for the "Magic Bullet." This approach has resulted in a flood of data, in response to which powerful statistical/data-driven/bioinformatics tools have been developed. These advances, however, have led to a development that is at the crux of the current scientific environment: the power and sophistication of these correlative methods has allowed "statistics" to evolve from a tool for the comparison of data sets to a means of using technology-generated insight as a substitute for knowledge. These purely associative relationships have become the basis for inferring causal mechanisms to be used in hypothesis formulation. But as stated above, the complexity and dimensionality of current biomedical problems, especially in terms of translating mechanisms across scales of organization, essentially precludes sole reliance on this approach. Based on Assertion 3, the situation can be described thusly: if it took a sophisticated data-mining, pattern-identification algorithm to identify correlative patterns from whence are abduced mechanistic hypotheses then attempting to construct traditional experiments to represent the system sufficiently with all its necessary richness and effectively test the multitude of hypotheses posed, is an intractable challenge. Currently, data is king, but data is not knowledge, and science is ultimately the advancement of knowledge. Correlative algorithms simply do not incorporate dynamic mechanistic relationships, and thus they cannot be used to suggest the outcome of any alteration of the system.

In the current, complex, data-rich environment, we assert that it is the step of hypothesis visualization and mechanistic causality evaluation that must be targeted for future technological augmentation. This step must be carried out in concert with, not in opposition to, the gathering of data and the analysis of these data using the types of data-driven modeling from which key insights amenable to dynamic mechanistic modeling can be extracted (Figure 5.1.1). The vast potential arising from advancements in computer technology must be utilized to perform "transscale" *in silico* experiments as a means of hypothesis checking to identify and clarify the dynamic consequences of a particular hypothesis. We have noted that all researchers construct and harbor mental models that form the intellectual basis of their research; these are often represented in a flow diagram of the type so ubiquitous to biomedical papers and presentations. However, these diagrams are static, insofar as they represent components, relationships, and mechanisms, but do not allow the consequences of these relationships to be visualized or evaluated. What is required is a procedure to bring these static diagrams "to life." This process has been termed executable biology [2], dynamic knowledge representation [3], or synthetic modeling and simulation [4]; we have covered this topic in

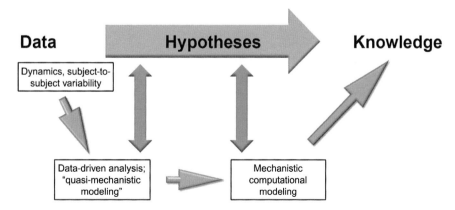

FIGURE 5.1.1   **The process from data to clinically actionable knowledge.** In order to obtain clinically actionable knowledge from high-content and high-throughput data, the data need to describe the dynamics of the biological process that underlies both the disease state and the basal healthy state, as well as accounting for patient-to-patient variability. Such data are amenable to data-driven modeling approaches, a suite of methods that can give quasi-mechanistic insights, which in and of themselves may serve as biomarkers as well as suggesting testable hypotheses. In order to increase the likelihood of deriving true clinically translational knowledge from data, however, hypotheses derived from data-driven modeling must be encoded as dynamic mechanistic computational models. The resultant mechanistic models could serve as the framework for *in silico* clinical trials, patient-specific diagnostics, or tools for rational design of drugs and devices. *Reprinted from [1].*

detail in Chapter 3.2 on *Dynamic Knowledge Representation. Note that this process is predicated upon the goal of sufficiency*: eventually the mental model will be broken, but knowledge advances via the process of breaking (à la Popperian Falsification), and, perhaps at some interval points, the mental model is sufficient and useful. Nowhere is this more important than in the case where the goal is trans-scale knowledge translation. Virtually by definition, the translational challenge precludes reliable human intuition; this fact makes the checking of multiscale translational mental models absolutely critical.

Dynamic mathematical modeling is the means for accomplishing this goal. We have given many examples in Section 4 of this book of the sorts of clinically relevant insights that can arise from the use of these types of models. We have also tried to make the case that this is not a new idea, not even for the biomedical research community. What is different about the approach we propose in this book? Our call for the integration of computational modeling at every key stage of the biomedical research process requires high-throughput modeling, and this, in turn, requires reducing the threshold for model construction. This goal cannot be achieved by placing priority on the development of highly detailed, high-resolution simulations of limited aspects of the pathophysiological process; i.e., a focus in producing quantitatively validated computational models of biological proxy models. Viewing the goal of computational modeling as the replication of biological proxy models (cell cultures, animal experiments, etc.) only recapitulates the fragmented continuum of the current biomedical research structure, and misses the power of using computational representation as a means of binding together disparate biological observations. These types of targeted simulations have their place (such as aiding in the process of refining our understanding of specific mechanisms), but a sole emphasis on the production of quantitative, "engineering grade" simulations misses the essential point needed for translational research: discovering and testing what we know, determining the expansiveness and applicability of what we know, and then building upon those hypotheses sufficient trust to move toward the development of clinical interventions. Thus, the issue of sufficiency is raised again: at what point is a simulation sufficient for high-throughput hypothesis evaluation? Fortunately, biology is structured around multiple scales of robustness and dynamic stability [5,6]. This architecture, in turn, suggests that the macro-level behaviors of biological systems are fairly resistant to micro-level fluctuations of underlying mechanisms. Therefore, assessments of macro-level behavior can be evaluated qualitatively with the goal of establishing *face validity* of a trans-scale mental model; i.e. establishing that the model behaves in a reasonable, plausible and recognizable way.

We assert that this process could take place in parallel, involving multiple interdisciplinary research teams, each with multiple hypotheses exploring potential solutions. This would be Translational Systems Biology implemented at a community-wide level, as multiple hypotheses are evaluated concurrently. Figure 5.1.2 demonstrates a schematic of what a computationally augmented, community-level knowledge ecology might look like [7]. At first, this prospect appears to be overwhelming, both in terms of implementation and analysis. The dilemma arises from the unavoidable tension between the quest for detail and the need to maintain scope: precision versus comprehension, quantitative versus qualitative, predictability versus ambiguity, brittle versus robust, and formal systems versus natural language. However, developments in computing technology, such as the types of automated model construction algorithms we have described in Chapter 4.5, combined with high-performance computing, could well provide precisely this capability. In short, formal hypothesis representation in computable form would finally allow biomedical researchers to "show their work."

It is important to note that the high-throughput hypothesis evaluation approach utilizes exactly the same scientific process that has been in practice for nearly 500 years. Our proposal within the framework of Translational Systems Biology only enhances the ability to carry out this tried-and-true process much more efficiently, rigorously, and rationally. Being able to implement such a strategy is no small task; perhaps it has not been done so far because it was technically impossible. We assert that this is no longer the case, and we have attempted, in this book, to outline a roadmap and highlight the toolsets available for this wholesale integration of dynamic computational modeling into the Scientific Cycle. The question is not "why"—we have highlighted the inability to translate a large percentage of promising basic science findings into clinically useful outputs—but rather "could it happen" and "what would change if it did happen"?

HIGH-THROUGHPUT DYNAMIC KNOWLEDGE REPRESENTATION: A COMMUNITY EFFORT

How would community-scale Translational Systems Biology be carried out? Circling back to our core interest in critical illness, let us first consider a hypothetical drug development scenario. To best understand the significance and potential impact of transforming the study of inflammation to a Translational Systems Biology model,

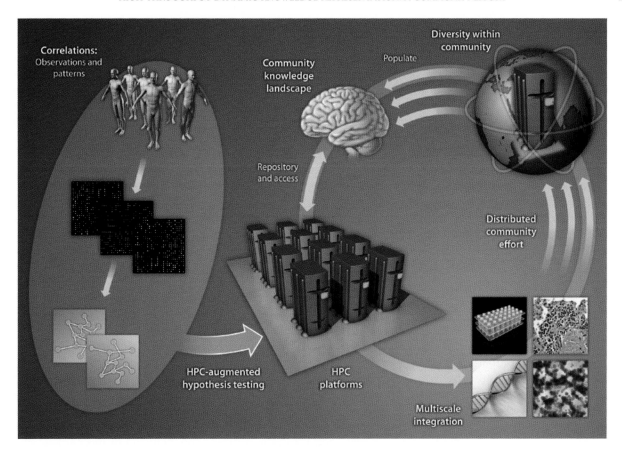

FIGURE 5.1.2 **Knowledge ecology for the future.** A proposed future structure of the biomedical research community, implementing the computationally augmented Scientific Cycle at a community-wide level. The key principle depicted here is the transparency in the representation, generation, and utilization of mechanistic knowledge as encapsulated in computational models. The freely and openly available biomedical knowledge allows for distributed investigation, aggregation of disparate knowledge, and the determination of the "fitness" of hypotheses based on their use and persistence within the research community. This research community structure incorporates evolutionary principles as applied to pieces of knowledge in a robust and scalable fashion. *Reprinted from [7].*

consider the following present-day scenario, which is typical of the fragmented nature of healthcare. In a biology department, basic researchers studying sepsis know every detail of a key receptor's signal transduction pathway for sensing bacterial products, but may not be familiar with the clinical manifestations of sepsis or the intricacies of randomized, placebo-controlled clinical trials because this information is more readily available to clinicians. In a pharmaceutical company, the emphasis is on the "druggability" of a promising receptor antagonist derived from this basic research, but there may be little appreciation for the role of this receptor in modulating inflammatory responses induced by agents other than one particular bacterial product that drives some part of the inflammatory response via the aforementioned receptor. At a major academic clinical center, physicians carry out trials of the receptor antagonist. They are limited by FDA regulations that combine all patients who exhibit any of a constellation of symptoms under the rubric of "sepsis." These clinicians discharge the roughly one-third of their patients who survive the 28-day period mandated by the Food and Drug Administration, but now chronic and rehabilitative care providers must deal with the increased risk for various complications that are the legacy of a bout with sepsis. All of these specialists are practicing the best care possible within their segment of the continuum of care from Preclinical studies→Clinical trials→In-hospital care→Chronic/rehabilitative care. However, all of these earnest efforts are being carried out with eye toward domain-specific optimization, but with little regard for the optimization of the process as a whole.

The foregoing highlights much of what, in our opinion, is wrong with the current biomedical research enterprise. What we do not mention is that if the indication was traumatic injury rather than sepsis, there would likely be no drug candidate at all to even discuss. How would a future in which Translational Systems Biology has been adopted look, and what impact would it have?

V. A NEW SCIENTIFIC CYCLE FOR TRANSLATIONAL RESEARCH AND HEALTH-CARE DELIVERY

SUCCESS THROUGH FAILURE

We foresee a future in which drugs are discovered by first modeling the disease on a computer, clinical trials of that drug are carried out on virtual patient cohorts generated from that same model before a single patient is enrolled, and where diagnosis of the efficacy of that drug in any individual patient is determined by a patient-specific variant of this very same model. Thus, the process of finding a new sepsis drug would include the workflow described in Section 3 of this book: dynamic knowledge representation using dynamic computational models to provide rational, model-driven design and testing of novel therapies; clinical trials that are first run *in silico*; inpatient care in which diagnosis is aided by mathematical models; and outpatient care plans prepared using model-driven decisions along the continuum of care. Within the framework, the continuum of care would be treated as an integrated whole, one in which biologically based mathematical models are used at every stage to optimize patient outcome. This is an ambitious undertaking, in which disease-specific, patient-specific, and treatment-specific elements would be unified using an iterative process of model building and validation. Perhaps, if such a workflow were to be adopted, we would be able to obtain drug candidates for traumatic injury, as well.

A key principle of the biotechnology and pharmaceutical industries is "fail early, fail fast, fail often", but this has not worked as intended, and has in many cases prolonged the process of drug discovery as well as making it more expensive [8,9]. The core principle that underlies our proposed approach might be termed "multistep falsifiability," in that the focus is not on the positive finding (which only suggests that a plausible solution has been found), but rather in the negative result, in the disproof of a particular hypothesis. Given the robustness inherent in the multiscale organization of biological systems, qualitative knowledge representation is likely to be sufficient for the goal of disproof. When implemented at the community level, and spanning the range from small basic science labs to large pharmaceutical companies, this approach has the potential to transform biomedical research. In this vision, the community knowledge landscape would consist of a series of *plausible* explanations, constantly being scrutinized via ongoing observation by the research community. The fitness of each hypothesis—and later, drug candidate or predicted outcome of an *in silico* clinical trial—would be determined by its ability to stand the scrutiny of comparison to real world observations. Those hypotheses that continue to be used will be those that survive; importantly, the reasons for survival or lack thereof would be explicit and transparent due to the inclusion of mechanistic dynamic modeling. In this fashion, the development of community knowledge—and, we would argue, new therapeutics and diagnostics—would follow an evolutionary paradigm, dependent upon diversity and selection. Representing knowledge in this fashion will in essence democratize the scientific process, where the fitness of an idea is not dependent upon its pedigree, but its success in surviving community scrutiny. Thus, by describing the process of scientific discovery formally we are able to apply technological augmentation where it is truly needed, and in so doing place the individual process into a community-wide context to provide a robust methodology for the advancement of science.

We emphasize that our goal is not the creation of "end-product models," but rather the integration of modeling into the research workflow and enhance the generation of clinically actionable scientific knowledge. Therefore, an iterative scheme that results in successive reduction in the set of possible explanations is one of the pragmatic goals of our computationally enhanced research strategy. Furthermore, as the scientific process has a cyclical structure, we focus on the use of iteration as a means of successive refinement and progressive reduction in the set of plausible hypotheses.

We envision this approach as proceeding in a multistep manner, though we note that the process is actually iterative and recursive, and so any given individual or group could enter this process at nearly any step. This overall schematic is displayed in Figure 5.1.3. First, we propose carrying out a series of experiments on clinically relevant animal models with well-defined, highly specific phenotypes to produce a repetitive sampling of target tissues over the course of disease progression. Concurrently, we propose the utilization of a combination of high-throughput technologies to generate a diversity of "reference" experimental data about the molecular components. Next, we envision organizing these data in an integrated knowledge base that could be accessed easily by the stakeholders, as well as funding and regulatory agencies. Next, we propose standardizing the process of design and development of the software architecture to provide an *in silico* experimental workflow. This workflow could include a "modeling sandbox" that augments hypothesis-driven investigation through semiautomated translation of evolving concepts and hypothesis into computational models for simulation and analysis. This step would provide an ensemble of predictive mathematical modeling methods for conceptual model verification, understanding the dynamic behaviors and functions of disease networks, investigating putative new molecular pathway interactions, analyzing potential effects of perturbations and interventions, and designing new biological experiments.

Next, we would propose the use of human cells to investigate and validate predicted molecular pathway mechanisms, trajectories, and outcomes from computationally evaluated hypotheses, perturbations, and interventions.

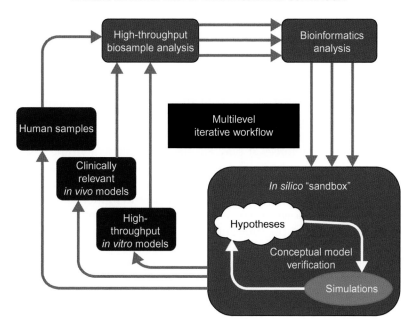

FIGURE 5.1.3 **Iterative multiscale workflow of Translational Systems Biology.** Translational Systems Biology utilizes basic scientific principles, manifest as the Scientific Cycle, as augmented by technology and operating at multiple complementary scales. The use of dynamic computational models within the "*in silico* sandbox" provides both a means of addressing the current process bottleneck in the data-rich imbalanced Scientific Cycle, as well as a means of integrating the experimental and observational output at different phases of the bio-investigatory process.

The use of human cells could provide a bridge from the molecular pathways discovered in our clinically relevant animal models to our eventual clinical human samples. These experiments would produce additional samples for the high-throughput technologies described above and comprise one layer of the multilayered iterative cycle of computational modeling, experimental validation, and data analysis.

This would progress to the design and execution of additional experiments on clinically relevant animal models with perturbations and interventions from computationally evaluated hypotheses. These experiments would produce additional samples for high-throughput technologies, as well as perform *in vivo* confirmation of disease progression and outcome from predictive mathematical models. This process comprises the second layer of the multilayered iterative cycle of computational modeling, experimental validation, and data analysis.

The next step would involve calibrating our predictive mathematical models to the high-throughput data measurements from normal and pathological human samples. Computational investigation of these models is expected to identify the clinically contextual dynamics and trajectories of particular disease states. Differences between model behavior that generates normal states versus model behavior that generates pathologic states would provide insights about the molecular pathway components and functions associated and potentially involved in pathogenesis.

Our sincere hope is that by presenting the case for this "better mousetrap" that involves the incorporation of computational modeling into the biomedical research enterprise, and especially for the fact that modeling adds relatively little cost and time overhead to the fragmented continuum as it exists today, that the various stakeholders—from basic researchers to industry to clinicians to funders of research and policymakers—would adopt this proposed approach.

A CASE FOR DISRUPTION OF THE FRAGMENTED CONTINUUM

As stated above, we are cognizant of the fact that the fragmented continuum of health-care delivery is, in and of itself a multiscale complex system that, to a great degree, evolved to generate the current incentive structure and propagates to maintain the *status quo*. Let us examine a hypothetical scenario in the setting of the pharmaceutical or biotechnology industry. Multiple group leaders are meeting in order to discuss current progress and possible new approaches. The Chief Science Officer is berating those gathered about the lack of promising new drug leads. There is mostly silence, since no one wants to be perceived as making excuses for failure. One person among those gathered is an advocate for computational modeling, possibly leads a modeling group within the company. This person considers the possibility of speaking up, eloquently, about how computational modeling could help lead to new drug candidates. However, it is difficult to make the case with few words, in a short time, in front of an audience that at best is uninformed and at worst is hostile. It is important to note that at that moment all those gathered are, in essence, *failing equally*. Indeed, speaking up and advocating computational modeling means adding cost (however small) to the overall operations of the pharmaceutical company, with little guarantee of a payoff (since, as we have described above, computational modeling is likeliest to be of most benefit when deployed throughout the process from preclinical studies to clinical trials and beyond). So, the likeliest outcome of this meeting is that *nothing at all will happen*.

Nothing will happen because all of the forces at play are stacked against anything disturbing the *status quo*: the company's management is typically conservative, the group leaders are politically savvy and worried about their jobs, the federal government does not want to interfere, academic scientists are themselves no more committed to incorporating modeling into the discovery process, and policymakers and funders are, to a very real degree, unaware of all of the above. As illustrated by the above scenario, those very same market forces and the self-sustaining holding pattern that characterize and drive the pharmaceutical industry in fact assure that, unless spurred to change little innovation will occur.

So how can this vicious cycle be stopped? We suggest that meaningful change will likely have to occur through disruption. The changes we propose could happen gradually, but a realistic assessment of the current biomedical enterprise, as we have detailed throughout this book, suggests that the current structure is too self-sustaining to make organic change unaided. The reality is that the holding pattern in which the entire biomedical enterprise currently finds itself is simply too difficult to break through. An enlightening parallel for the powerful effects of disruption is the technology industry. In this setting, disruption forced upon a company subsequent to a very specific type of takeover, which we discuss below, can in theory happen very quickly. If we look at recent history in the technology industry, established players in the technology space have fallen (or are in imminent danger of falling) from their lofty perches almost overnight.

A key characteristic of the technology industry is that disruption is incentivized. Leading companies in the technology industry are accustomed to a different dynamic, in which large companies push other large companies to change, and also in which small companies can challenge large companies because the barriers to entry are relatively low. The key observation is that those engaged in that industry truly understand the "adapt or die" ethic. So, they are willing to take a chance, and disparage those that are not equally willing. We would argue that the end user benefits, even though orphan products and fragmentation may be a byproduct of this constant, ongoing disruption. It is quite interesting that the leaders of tech have already begun to tiptoe toward the health-care market, though currently focused on "wearables" and health/fitness monitoring. But they have not—yet—acquired a biotechnology or pharmaceutical company…

Imagine, then, a scenario in which a major player in the tech field buys a solid biotechnology company whose shares are undervalued. This acquisition will spur many hurried meetings at competing biotech and pharmaceutical companies, because tech companies are known for disruption, for creating entirely new fields, and for moving at a very fast pace. And if this new company is successful… the process would snowball. Shareholders of competing companies would demand a similar process. Perhaps the federal government would notice as well, and begin to enforce its own "Critical Path" guidelines [9]. This may be a just a figment of our imagination, a fantasy. However, it is completely possible given current trends and market forces, and the fact that change in all aspects of society is happening faster and faster, catalyzed by key events.

Disruption aimed at incorporating computational modeling into the current biomedical research process may well occur via other means, for example, due to a government mandate driven by considerations other than those that perpetuate the current *status quo*. The key point that must be recognized is that, as with all processes subject to selection, there is going to be a "fitness function"; some means by which "success" is judged and is the direct product of the incentive structure present. Currently, the context in which biomedical research "success" is judged is too narrow and short-sighted: it is the granting cycle, or the length of an academic appointment, or the interval of a shareholder's meeting. Paradoxically, the length of the drug development pipeline is too long: years and perhaps even decades pass between the time of discovery of a promising drug candidate to the time it can find its way to the bedside. The discordance between the short-term socio-professional-economic incentive structure and the time it takes to actually accomplish the ostensible goal of the biomedical research enterprise only reinforces the acceptance of long-term failure for short-term gain.

Translational Systems Biology is intended to bring these two disparate timelines closer into concordance. The operational research methods involving computational modeling may not at first accelerate the time it takes to get a drug finally approved, but it will increase the efficiency of that pipeline and offers the only pathway forward toward potentially speeding the drug approval process in a safely meaningful way. The cultural transformation that needs to occur, founded upon the paradoxical ability of the technology industry to take a longer view within a rapidly evolving sector, will provide a restructuring of the incentive structure that gives the research methods of Translational Systems Biology the chance to meet their fruition, given the recognition that what is being implemented is a robust and scalable "solution technology" as opposed to an all-or-nothing product. Participants at both ends of the spectrum can derive confidence in the "return to the future" offered by Translational Systems Biology: it represents an organic, technology-aided growth of the proven Scientific Method from which so much of our current civilization derives. We suggest that Translational Systems Biology offers a pathway toward the convergence of these two incentive-dependent "fitness functions" that govern the current practice of biomedical research, and in so doing, can release us from the Translational Dilemma we find ourselves in today. We can only hope that a rational future awaits.

References

[1] An G, Nieman G, Vodovotz Y. Computational and systems biology in trauma and sepsis: current state and future perspectives. Int J Burns Trauma 2012;2(1):1–10. PMID: 22928162.

[2] Fisher J, Henzinger TA. Executable cell biology. Nat Biotechnol 2007;25(11):1239–49.

[3] An G. Introduction of an agent-based multi-scale modular architecture for dynamic knowledge representation of acute inflammation. Theor Biol Med Model 2008;5(1):11.

[4] Hunt CA, Ropella GE, Lam TN, Tang J, Kim SH, Engelberg JA, et al. At the biological modeling and simulation frontier. Pharm Res 2009;26:2369–400.

[5] Doyle J, Csete M. Rules of engagement. Nature 2007;446(7138):860.

[6] Csete ME, Doyle JC. Reverse engineering of biological complexity. Science 2002;295(5560):1664–9.

[7] An G. Closing the scientific loop: bridging correlation and causality in the petaflop age. Sci Transl Med 2010;2(41):41ps34.http://dx.doi.org/doi:10.1126/scitranslmed.3000390 PMID: 20650869.

[8] Bonabeau E, Bodick N, Armstrong RW. A more rational approach to new-product development. Harv Bus Rev 2008;86(3):96–102. 134.

[9] U.S. Food and Drug Administration. Innovation or stagnation: challenge and opportunity on the critical path to new medical products; March 2004.

Index

Note: Page numbers followed by "*f*" refer to figures.

CPI Antony Rowe
Eastbourne, UK
February 01, 2015